T0134453

Progress in IS

More information about this series at http://www.springer.com/series/10440

Benoît Otjacques · Patrik Hitzelberger
Stefan Naumann · Volker Wohlgemuth
Editors

From Science to Society

New Trends in Environmental Informatics

Editors
Benoît Otjacques
Environmental Research and Innovation
Luxembourg Institute of Science
  and Technology
Esch/Alzette
Luxembourg

Patrik Hitzelberger
Environmental Research and Innovation
Luxembourg Institute of Science
  and Technology
Esch/Alzette
Luxembourg

Stefan Naumann
Institut für Softwaresysteme in Wirtschaft,
  Umwelt und Verwaltung
Hochschule Trier - Umwelt-Campus
  Birkenfeld
Birkenfeld
Germany

Volker Wohlgemuth
Environmental Informatics
University of Applied Sciences
Berlin
Germany

ISSN 2196-8705   ISSN 2196-8713   (electronic)
Progress in IS
ISBN 978-3-319-88080-8   ISBN 978-3-319-65687-8   (eBook)
DOI 10.1007/978-3-319-65687-8

Printed on acid-free paper

This Springer imprint is published by Springer Nature
The registered company is Springer International Publishing AG
The registered company address is: Gewerbestrasse 11, 6330 Cham, Switzerland

Preface

This book presents the main research results of the 31st edition of the long-standing and established international and interdisciplinary conference series on environmental information and communication technologies (EnviroInfo 2017).

The conference was held on 13–15 September 2017 at the Neumünster Abbey in Luxembourg, organized by the Luxembourg Institute of Science and Technology (LIST), under the patronage of the Technical Committee on Environmental Informatics of the German Informatics Society.

The tag line of this year's conference was "From Science to Society: The Bridge provided by Environmental Informatics". It underlines that Environmental Informatics has its roots in applied research and that it is not an ivory tower discipline. Environmental research at the local organizer LIST, a research and technology organization, is a clear example of this focus on doing research that has a high impact on the society. At the same time, the location Luxembourg truly exemplifies some of the complex research topics for Environmental Informatics today. For example, due to its size and geographical situation, the country is specifically challenged by environmental issues that cross borders such as the impact of environmental disasters, emissions caused by the cross-border mobility due to the international workforce in the country, or the interoperability of environmental decision systems in general.

The articles in this book give some innovative answers to these and other questions that are relevant for Environmental Informatics research, such as semantic interoperability modelling used for disaster management, sustainable mobility solutions, approaches for energy-aware software-engineering, and land-use planning.

The editors would like to thank all the contributors to the conference and these conference proceedings. Special thanks also go to the members of the programme and organizing committees. In particular, we would like to thank all those involved at the local organizer, LIST. Last, but not least, a warm thank you to our sponsors that supported the conference.

Belvaux, Luxembourg   Benoît Otjacques
Belvaux, Luxembourg   Patrik Hitzelberger
Birkenfeld, Germany   Stefan Naumann
Berlin, Germany   Volker Wohlgemuth
June 2017

EnviroInfo 2017 Organizers

Chairs

Dr. Benoît Otjacques, Luxembourg Institute of Science and Technology, Esch/Alzette, Luxembourg
Prof. Dr. Stefan Naumann, Hochschule Trier—Umwelt-Campus Birkenfeld, Trier, Germany

Programm Committee

Arndt, Hans-Knud
Bartoszczuk, Pawel
Bunse, Christian
Del Frate, Fabio
Dieudé-Fauvel, Emilie
Düpmeier, Clemens
Fischer-Stabel, Peter
Fishbain, Barak
Förster, Anna
Fuchs-Kittowski, Frank
Funk, Burkhardt
Geiger, Werner
Greve, Klaus
Heikkurinen, Matti
Hilty, Lorenz M.
Hitzelberger, Patrik
Hönig, Timo
Jensen, Stefan
Johann, Timo

Karatzas, Kostas
Kern, Eva
Klafft, Michael
Knetsch, Gerlinde
Knol, Onno
Kremers, Horst
MacDonell, Margaret
Marx Gómez, Jorge
Mattern, Kati
Möller, Andreas
Müller, Berit
Naumann, Stefan
Niemeyer, Peter
Niska, Harri
Ortleb, Heidrun
Otjaques, Benoît
Page, Bernd
Parisot, Olivier
Pattinson, Colin
Pillmann, Werner
Rapp, Barbara
Schade, Sven
Schaldach, Rüdiger
Schöner, Dominik
Schreiber, Martin
Simon, Karl-Heinz
Sonnenschein, Michael
Susini, Alberto
Tamisier, Thomas
Thimm, Heiko Henning
Voigt, Kristina
Wagner vom Berg, Benjamin
Willenbacher, Martina
Winter, Andreas
Wittmann, Jochen
Wohlgemuth, Volker

Contents

Part I
Applications of Geographical Information Systems and Disaster Management

Forecasting the Spatial Distribution of Buildings that Will Remain in the Future

Toshihiro Osaragi and Maki Kishimoto

Abstract Changes in land use in established city areas generally arise when older style buildings are demolished and replaced by contemporary buildings. The direction and the speed of land use change are dependent on the possibility and probability that buildings will be demolished or will remain in the future. Hence, various studies about the life span of buildings have been carried out, and proposed statistical models that could estimate the value of probability that buildings would be demolished or would remain in a specific time interval, based on the age of buildings. However, in general it is not easy to acquire the necessary information about the age of buildings. In order to extend the application of the proposed model to these cases, we propose a method for estimating the number of buildings that will remain in the future when data about the age of buildings cannot be obtained.

Keywords Interval probability function of remainder · Age of building · Life span · Disaster prevention planning · Land use model · Tokyo

1 Introduction

A basic unit of land use change is a building lot owned by a landowner or business proprietor. Land use changes generally arise when older style buildings are demolished and replaced by contemporary buildings. Such phenomena are often observed in established high-density areas. Hence the direction and the speed of land use change are dependent on the possibility and probability that buildings will be demolished or will remain (Osaragi and Nishimatsu 2013). This is quite different

T. Osaragi (✉) · M. Kishimoto
Department of Architecture and Building Engineering, School of Environment and Society, Tokyo Institute of Technology, Tokyo, Japan
e-mail: osaragi.t.aa@m.titech.ac.jp

M. Kishimoto
e-mail: kishimoto.m.ac@m.titech.ac.jp

© Springer International Publishing AG 2018
B. Otjacques et al. (eds.), *From Science to Society*, Progress in IS,
DOI 10.1007/978-3-319-65687-8_1

from land use change in suburban areas, where a major concern is "change of land-use classification, for example, change from field/forest to residential area".

In respect of the field of disaster prevention planning, conversion of building structures or materials to an incombustible state is one of the most important issues (Osaragi 2013). Although research regarding the provision of fire-fighting services is of relevance to these disasters (Adrian et al. 1983, Richard et al. 1990; Wallace 1993), we should discuss fire prevention rather than control in order to comprehensively address the problem of multiple uncontrolled fires following a large earthquake (Wallace 1991). The process of conversion of cities to an incombustible state has not thus far been strongly promoted, because of the large amount of resources necessary for rebuilding. Conversion of buildings to an incombustible state is generally achieved when older style buildings are demolished and replaced with contemporary buildings. The speed at which building structures or materials may be changed is thus dependent also on the possibility and probability that buildings will be demolished.

Various studies about the life span of buildings have been carried out (Kaplan and Meier 1958), since the life span of buildings is one of the basic concerns in various fields including urban economics and management, land use, and disaster prevention planning. Reliability theory has been applied in many studies, and they, in turn, proposed methods for assessing the probability that buildings will be demolished or remain in the future (Komatsu 1992; Osaragi et al. 2002). However, most of these studies have considered the case that the life span of buildings is a simple, single-variable function dependent only on the age of the building. Given this background, Osaragi (2004) proposed a statistical model that can evaluate characteristics of buildings and location, which in turn affect the life span of buildings. In particular, he examined characteristics of buildings (age of building, construction materials, building type, etc.) and characteristics of locations (land use zoning, building area to plot ratio, accessibility to railway stations, etc.), and evaluated how these factors affected the life span of buildings. Furthermore, he proposed a mathematical model, referred to as the "interval probability function of remainder", that can stably estimate the probability that buildings will be demolished or will remain within a time interval (Osaragi 2005). That is, if information about the age of each building is acquired, the demolished/remaining probability can be estimated and applied in a variety of micro-simulation models.

However, in the general use of the model, the age of each building must be known in order to estimate the value of the probability. This requirement is an impediment to the application of the proposed model to actual urban data. In this article, in order to modify this disadvantage and extend applications of the model, an alternative method of the model is proposed using the concept of the "average decrepitude of buildings", which can be estimated from time series digital maps of cities. The average decrepitude of buildings can be considered as a quasi-average-age of buildings in an area, which can be calculated from the number of buildings that remain in a time interval. The utility and the efficiency of the alternative method are examined by consideration of numerical examples based on actual data taken from densely built-up areas of the Tokyo metropolitan area.

2 Modeling of the Probability Function of Remainder

The age of buildings is denoted by t $(t \geq 0)$, and the category of buildings or locations (construction materials, building type, land use zoning, building area to plot ratio, accessibility to railway station, etc.) is denoted by j $(j = 1, 2, \ldots, m)$. The probability that a building of category j exists at age t is expressed by $P_j(t)$. This function is labeled the "probability function of remainder". It is assumed that derivative of $P_j(t)$ with respect to age t can be obtained. Moreover, "the density function of life span" is expressed by $f_j(t)$, and "the demolition function" which expresses the ratio of buildings being demolished is denoted by $h_j(t)$. The following relations exist between the above variables,

$$f_j(t) = \frac{-dP_j(t)}{dt},$$ (1)

$$h_j(t) = \frac{f_j(t)}{P_j(t)}.$$ (2)

On the substitution of the expression for $f_j(t)$ derived from Eq. (2) for $f_j(t)$ in Eq. (1), we obtain

$$P_j(t) = \exp\left[-\int_0^t h_j(x)dx\right].$$ (3)

Although it is important to investigate the characteristics of demolition function $h_j(t)$, for the present purposes it is assumed that $h_j(t)$ can be expressed by the following equation using Taylor's expansion,

$$h_j(t) = \sum_{k=0}^{K-1} \frac{h_j^{(k)}(a)}{k!}(t-a)^k,$$ (4)

where $h_j^{(k)}(a)$ is the derivative of $h_j(t)$ with respect to time at the age $t = a$, and K is a constant sufficiently large to ensure a satisfactory approximation. Expressing the demolition function $h_j(t)$ as a polynomial of age t in this way, Eq. (3) can be rewritten as follows,

$$P_j(t) = \exp\left[-\sum_{k=1}^{K} a_{jk}t^k\right],$$ (5)

where a_{jk} is a constant coefficient corresponding to the coefficients k of the powers of t in Eq. (4). Specifically, the demolition function, $h_j(t)$, can be expressed using the coefficients a_{jk} as follows:

$$h_j(t) = \sum_{k=1}^{K} k a_{jk} t^{k-1}. \tag{6}$$

The probability that a building, which has survived to age t, continues to survive to age $t + \Delta t$ ($\Delta t > 0$) is expressed by $P_j(t + \Delta t|t)$. This function is hereafter called "the interval probability function of remainder". Since $P_j(t + \Delta t|t)$ is a conditional probability of $P_j(t)$, it has the following relationship with the probability function of remainder, $P_j(t)$,

$$P_j(t + \Delta t|t) = \frac{P_j(t + \Delta t)}{P_j(t)}$$
$$= \exp\left[-\sum_{k=1}^{K} a_{jk}\{(t + \Delta t)^k - t^k\}\right]. \tag{7}$$

Thus, the values of $P_j(t)$ and $P_j(t + \Delta t|t)$ can be obtained and applied to a variety of simulation models, if the coefficients a_{jk} are estimated.

The values of a_{jk} can be estimated as maximum likelihood estimators by applying the undetermined multipliers method of Lagrange. Due to the limitation of pages, we omit the detailed description for the estimation method, which shows that the Lagrange function is a convex function of the coefficients a_{jk}. That means, a local minimum solution, which is a typical problem in optimization by the Hill-climbing method, does not exist for the present problem.

3 Methods for Estimating the Number of Buildings that Will Remain

We describe four methods, shown in Fig. 1, for predicting the number of buildings that will remain in through a time interval. First, if information about ages of buildings is acquired, the following two methods are available, which use the probability function of remainder directly.

Method (1): Substituting the age of each building for the variable t in $P_j(t + \Delta t|t)$, the value of the interval probability of remainder is then predicted for each building. This method thus requires detailed information about building ages to be effective in predicting the demolished/remaining status of each building, one by one.

Method (2): Substituting the mean value of buildings age $\bar{t}$ in a study area for the variable t in $P_j(t + \Delta t|t)$, the average value of the interval probability function of remainder is then derived. This method can be used in situations where not all of the ages of the buildings are known, but the mean value of the ages of buildings in the study area is known. It differs from method (1) in that the value of the interval probability function of remainder is therefore assumed to be the same constant value for all buildings of the same type in the area. Accuracy is thus expected to

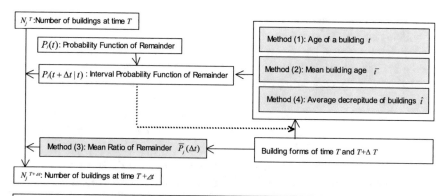

N_j^T: Number of buildings at time T

$P_j(t)$: Probability Function of Remainder

$P_j(t+\Delta t \,|\, t)$: Interval Probability Function of Remainder

Method (1): Age of a building t

Method (2): Mean building age $\bar{t}$

Method (4): Average decrepitude of buildings $\hat{t}$

Method (3): Mean Ratio of Remainder $\overline{P}_j(\Delta t)$

Building forms of time T and $T+\Delta T$

$N_j^{T+\Delta t}$: Number of buildings at time $T+\Delta t$

Method (1): Age of each building t
Substituting the age of a building for the variable t
in $P_j(t+\Delta t|t)$, the value of the interval probability of remainder
is derived for each building.

$$N_j^{T+\Delta t} = \sum_t N_j^T(t) P_j(t+\Delta t \,|\, t)$$

$N_j^T(t)$: Number of buildings of age t, type j at time T

Method (2): Mean building age $\bar{t}$
Substituting the mean value $\bar{t}$ of building age t in an analysis
area for the variable t in $P_j(t+\Delta t|t)$, the average value of the
interval probability of remainder is derived.

$$N_j^{T+\Delta t} = N_j^T P_j(\bar{t}+\Delta t \,|\, \bar{t})$$

Method (3): Mean ratio of remainder $\overline{P}_j(\Delta t)$
Comparing building forms of time T and time $T+\Delta T$, the values
of N_j^T and $N_j^{T+\Delta t}$ are counted and the value of the mean ratio
of remainder $\overline{P}_j(\Delta T)$ of term ΔT is derived. The value of
$\overline{P}_j(\Delta t)$ of term Δt can be estimated by calculating the $\Delta t / \Delta T$
power of the value $\overline{P}_j(\Delta T)$ is calculated.

$$N_j^{T+\Delta t} = N_j^T \overline{P}_j(\Delta t)$$

$$\overline{P}_j(\Delta t) = \left(\frac{N_j^{T+\Delta T}}{N_j^T} \right)^{\frac{\Delta t}{\Delta T}}$$

Method (4): Average decrepitude of buildings $\hat{t}$
Substituting the value of $\overline{P}_j(\Delta T)$ for the inverse function of $P_j(t+\Delta T|t)$,
the average decrepitude of building $\hat{t}$ at time T is estimated.
Furthermore, upon substituting this value $\hat{t}$ for the variable t in $P_j(t+\Delta t|t)$,
the value of the interval probability of remainder in term Δt
can be obtained.

$$N_j^{T+\Delta t} = N_j^T P_j(\hat{t}+\Delta t \,|\, \hat{t})$$

\<Method for estimating the interval probability of remainder\>

Probability that a building of age t_0 will continue to remain for
Δt can be estimated by substituting the age t_0 into the interval
probability function of remainder.

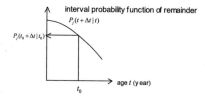

\<Method for estimating the average decrepitude of buildings\>

Estimate the mean ratio of remainder by using two maps of
building forms at time T and $T+\Delta T$. The average decrepitude of
buildings is estimated by substituting the mean ratio into the
inverse function of interval probability of remainder.

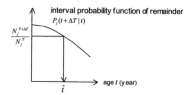

Fig. 1 Methods for estimating the number of buildings that will remain

diminish when the actual age of a building differs significantly from the mean value of building age in the study area.

If no information about ages of buildings is available, the following two methods can be applied by using digital maps of building forms at the two points in time.

Method (3): By comparing building forms at time T and time $T + \Delta T$ on geographic information systems (GIS), we can extract data for the number of buildings in existence at time T, N_j^T, and at time $T + \Delta T$, $N_j^{T+\Delta T}$. From these two values, the value of the mean ratio of remainder $\bar{P}_j(\Delta T)$ of term ΔT can be derived. Furthermore, the value of the mean ratio of remainder $\bar{P}_j(\Delta t)$ of term Δt can also be estimated by calculating the $\Delta t / \Delta T$ power of the value $\bar{P}_j(\Delta T)$. However, since the value of the mean ratio of remainder is assumed constant across any given interval, this method is not therefore considered to be suitable for long-term predictions.

Method (4): Substituting the value of $\bar{P}_j(\Delta T)$ for the inverse function of $P_j(t + \Delta t | t)$, the average decrepitude of building $\hat{t}$ at time T can be estimated. The average decrepitude of buildings $\hat{t}$ is considered as a quasi-mean-age of buildings in the area. Note that the value $\hat{t}$ may be larger or smaller than the actual mean age of the building. Substituting this value $\hat{t}$ for the variable t of $P_j(t + \Delta t | t)$, the value of the interval probability function of remainder in term Δt will be obtained.

4 Validation of Methods

Using the data for Mitaka city in Tokyo, the number of buildings that remain in given specific time intervals is estimated using the four methods, and their accuracy is examined. The interval probability function of remainder according to building type is constructed using data from 1994 and 1996. Then, the number of buildings existing in 1994 that are expected to remain 4 years later (in 1998), is estimated by the four methods. Estimated values according to the building type for each town (the number of towns is 11) are compared with the actual observed values. Here, the accuracy of the estimates is evaluated by the following absolute error rates,

$$e_i = \frac{1}{M} \sum_{j=1}^{M} \left| \frac{N_{ij} - \hat{N}_{ij}}{N_{ij}} \right| \times 100 \ [\%], \tag{8}$$

where a suffix j ($j = 1, 2, \ldots, M$) denotes a building type and a suffix i denotes a town ID number. The value N_{ij} is the actual observed value, while $\hat{N}_{ij}$ is the corresponding estimated value.

The absolute error rate of the number of remaining buildings estimated by each method is shown in Fig. 2. For "commercial buildings", the estimation accuracy of method (1) is superior to that of the other methods. Before performing this analysis, we expected that method (1), using the age of each building, would offer the highest stability and accuracy for all building types. However, the results show that these

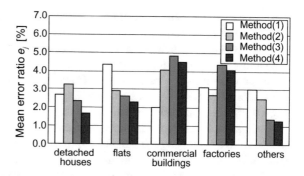

Fig. 2 Accuracy of estimates obtained by the four methods

expectations are not fulfilled. With this method, the same interval probability function of remainder is assumed across the whole area for buildings of the same type. In a small area like a town, however, the influence of the regional gap of demolition characteristics should not be ignored. Therefore, the estimation accuracy is not as high as was expected, even if detailed information of buildings' ages is used in the estimation. That is, we can conclude that the study area should be divided into sub areas in order to estimate the interval probability function of remainder more accurately if the demolition characteristics of building type are strongly dependent on local characteristics.

According to the results of "detached houses", "flats", and "others", the estimation accuracy of method (4) is higher than methods (1) and (2). As described in the above paragraph, the average decrepitude of buildings in each area used in method (4) reflects the demolition characteristics in the object area. Therefore, even if an area includes many comparatively new buildings, in practice, if the rate of demolition is high in the area, the degree of demolition might also be relatively high. Consequently, even if the same interval probability function of remainder is used, the estimate of method (4) may be of higher accuracy than methods (1) or (2), which do not incorporate the local characteristics of locations.

Next, the correspondence between the estimated values of the number of demolition buildings and the actual observed values are examined. The results are shown in Fig. 3. Each point in Fig. 3 corresponds to one town. Although some

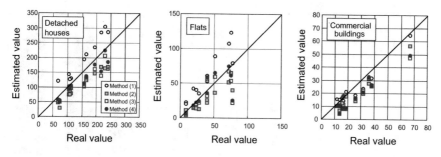

Fig. 3 Accuracy of estimates of the number of buildings demolished obtained by the four methods (Each point corresponds to a single towns in Mitaka city)

towns show large estimation errors, overall the estimated values and the actual observed values correspond well.

Next, we consider here the flexibility of application of the method of the interval probability function of remainder. The number of remaining buildings in Tokyo's Setagaya ward is estimated by method (4) using the interval probability function of remainder of Mitaka city. A corresponding estimate is also derived from method (3), and then the estimation accuracies of these two methods are compared. That is, the estimation accuracy is examined for the case when the ages of the buildings are unknown. By overlaying the building forms of 1991 and 1996 using the Tokyo GIS data, information of whether buildings are demolished or remain is obtained.

The number of buildings of remaining for 10 years (i.e. to 2001) is estimated, considering 1991 as the initial year. Specifically, since the mean ratio of remainder for 10 years is required to apply method (3), the squared value of the mean ratio of remainder for 5 years is estimated. In the process of applying method (4), first, the average decrepitude of buildings $\hat{t}$ in 1991 for each town is estimated by using the interval probability function of remainder. By substituting this value $\hat{t}$ for the interval probability function of remainder $P_j(t+10|t)$, the probability that each building will continue to remain after 10 years is estimated. The rate of the mean absolute error is calculated by Eq. (3), and the results are shown in Fig. 4. It is shown that the rate of the mean error of method (4) is smaller than that of method (3). For detached houses, flats, and commercial buildings, the rate of the mean error is around 5 percent or less. Since the study area is different from the area from which the parameters are estimated, the results are inferior to those shown in Fig. 2. However, the results shown in Fig. 4 are sufficiently accurate for practical purposes. For longer estimations, estimation accuracy will diminish, since method (3) assumes constancy of the mean ratio of remainder. On the other hand, since method (4) uses the average decrepitude of buildings, it is possible to describe the probability of remaining precisely, using an expression that varies with time. Therefore, method (4) shows a higher accuracy than method (3).

Next, in order to examine the estimation accuracy of method (4), the actual and estimated values of the number of demolition buildings for each town are compared (the number of towns in Setagaya ward is 62). The results are shown in Fig. 5. For

Fig. 4 Accuracy of estimates of the number of buildings that will remain obtained by methods (3) and (4) (Setagaya Ward)

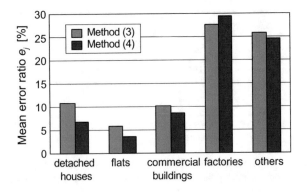

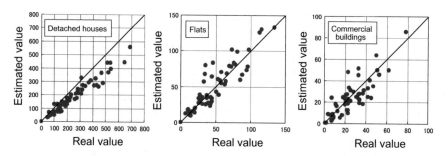

Fig. 5 Accuracy of estimates of the number of buildings demolished obtained by method (4) (Each point corresponds to a town in Setagaya Ward)

the case of detached houses, since the actual values of the demolition rate in 1991–1996 are lower than those of 1997–2001, the estimated values are a little small in many towns. We can say, however, that overall there is good agreement between the estimated and actual values.

5 Summary and Conclusions

In this research, a method for extending the application of the interval probability function of remainder is proposed, in order to facilitate the incorporation of this function into micro-simulation models, such as land use models and disaster prevention planning for established city areas. In particular, it is shown that the average decrepitude of buildings can be calculated if the data of building forms at two points in time can be obtained, even if information about the age of buildings is not available. Moreover it is shown that the number of remaining buildings can be estimated to a high degree of accuracy (the rate of error is about 1–4%) in a comparatively small spatial unit like a town. Furthermore, by using the interval probability function of remainder and the average decrepitude of buildings, estimations of the number of demolished/remaining buildings in the future can be achieved to a high accuracy.

References

Adrian C (1983) Analysing service equity: fire services in Sydney. Environ Plann A 15:1083–1100

Kaplan EL, Meier P (1958) Nonparametric estimation from incomplete observations. J Am Stat Assoc 53:457–481

Komatsu Y (1992) Some theoretical studies on making a life table of buildings. J Arch Plann Environ Eng 439:91–99 (in Japanese)

Osaragi T, Shimizu T (2002) Related factors to demolishing of buildings and estimation of subsistence probability. J Arch Plann Environ Eng 560:201–206 (in Japanese)

Osaragi T (2004) Factors leading to buildings being demolished and probability of remainder. In: Proceedings of the sustainable city III. WIT Press, pp 325–334

Osaragi T (2005) The life span of buildings and the conversion of cities to an incombustible state. In: Proceedings of safety and security engineering IV. WIT Press, pp 495–504

Osaragi T (2013) Towards an incombustible city: building reconstruction in potential and probable fireproofing of urban lots. In: Georisk: assessment and management of risk for engineered systems and geohazards. Springer. doi:10.1080/17499518.2013.769821

Osaragi T, Nishimatsu T (2013) A model of land use change in city areas based on the conversion of unit lots. In: Planning support systems for sustainable urban development. Lecture notes in geoinformation and cartography, vol 195, pp 31–49

Richard D, Beguin H, Peeters D (1990) The location of fire stations in a rural environment: a case study. Environ Plann A 22:39–52

Wallace R (1991) A stochastic model of the propagation of local fire fronts in New York City: implications for public policy. Environ Plann A 23:651–662

Wallace R (1993) Recurrent collapse of the fire service in New York City: the failure of paramilitary systems as a phase change. Environ Plann A 25:233–244

Research on the Potential Environmental Zonation of Red Flesh Dragon Fruit in Vinh Phuc Province

Minh Nhat Thi Doan, Cong Thien Dao, Nam Ta Nguyen, Hang Thanh Thi Nguyen, Hang Le Thi Tran, Son Thanh Le and Manh Van Vu

Abstract Red flesh dragon fruit (Hylocereus polyrhizus) or pitaya is well-known as an excellent source of antioxidants and high content of nutrients. The Red flesh dragon fruit is a hybrid of the white flesh dragon fruit Hylocereus undatus and a native Hylocereus costaricensis from Middle and North America. It brings high economic efficiency and currently being developed in the province of Vinh Phuc. The main objective of the study is to create a model, which help to determine the suitability in developing dragon fruit cultivation in Vinh Phuc province in order to support local agricultural and land use planning. The model is built based on the application of Geographic Information System (GIS), Kriging interpolation method, Multi-Criteria Decision Analysis (MCDA) method, Fuzzy Set Theory and Analytic Hierarchy Process (AHP) to evaluate and map the area suitable zonation for farming.

Keywords Hylocereus polyrhizus · GIS · MCDA · AHP

M.N.T. Doan (✉) · C.T. Dao · N.T. Nguyen · H.T.T. Nguyen · H.L.T. Tran · S.T. Le · M.V. Vu
VNU University of Science, 334 Nguyen Trai, Thanh Xuan, Hanoi, Vietnam
e-mail: dtnminh.2410@gmail.com

C.T. Dao
e-mail: dtcong2104@gmail.com

N.T. Nguyen
e-mail: namnguyenkhtn@gmail.com

H.T.T. Nguyen
e-mail: thanhhang0283@gmail.com

H.L.T. Tran
e-mail: hangtrancoste@gmail.com

M.V. Vu
e-mail: vuvanmanh@hus.edu.vn

© Springer International Publishing AG 2018
B. Otjacques et al. (eds.), *From Science to Society*, Progress in IS,
DOI 10.1007/978-3-319-65687-8_2

1 Introduction

Vinh Phuc is located in Red River Delta. It is one of the seven provinces of the Northern key economic zone. Vinh Phuc province is bordered to the North by Thai Nguyen and Tuyen Quang, to the west by Phu Tho, to the south by Hanoi, to the east by 2 districts: Soc Son and Dong Anh of Hanoi city. The province has 9 administrative units: Vinh Yen City, Phuc Yen town and the districts: Binh Xuyen, Lap Thach, Song Lo, Tam Duong, Tam Đao, Vinh Tuong, Yen Lac (Fig. 1).

Vinh Phuc is one of the provinces that have the highest level of GDP growth rate of the entire country. In order to ensure the sustainable development, the province has identified the direction of developing the service industry to re-invest in agriculture, mobilizing investment capital for production development and construction of agricultural infrastructure (Vinh Phuc Provincial People's Committee 2011). In planning the agricultural development of the province, the red flesh dragon fruit was added to the list of crops with high economic value. However, until now, this crop has not been studied for the zonation developed in accordance with the environmental conditions of the province.

Red flesh dragon fruit is hybrid between white flesh dragon fruit Hylocereus undatus and Hylocereus costaricensis, a native breed originated in Central and North America. Red flesh dragon fruit is one of the fruit plants which bring high economic efficiency in both developed North and South Vietnam, most concentrated in Binh Thuan. Dragon fruit can grow and develop in different soil types such

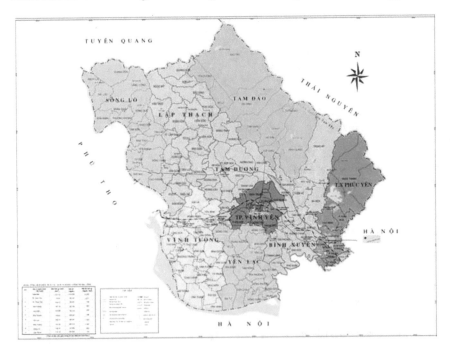

Fig. 1 Vinh Phuc administrative map

as sandy soil, gray soil, alkaline soil to alluvial soil, red soil bazan, soil. The temperature suitable for the red dragon fruit growth and development ranges from 15 to 35 °C (Van Ke 1998), in which the most appropriate temperature ranges from 21 to 26 °C (Liaotrakoon 2013). The weather conditions of light frost, even for a short time will also affect red dragon fruit growth. The demand of precipitation for crops ranges from 800 to 2000 mm/year (Van Ke 1998). A good drainage system can reduce the influence of precipitation on the flowering, fruiting of trees. The red dragon fruit is photophilic plants, suitable for cultivation in areas where there are lots of light. In addition, it can be planted in the low slope area in (Truong Thi Dep 2000).

2 Methodology

The research applied a series of criteria in order to find potential development of the red dragon fruit tree planting. The criteria include temperature, precipitation, distance to the irrigation source, distance to roads, slope.

2.1 Impact Factors for the Suitability of Growing Dragon Fruit

Temperature

Temperature is a factor that greatly affects plant growth. Plants can grow in a wide temperature range, but different types of plants have different minimum and maximum temperature points. The minimum temperature and the maximum temperature for the growth of the plant are the temperature points at which the plant stops growing. Limits of temperatures growth change with the adaptation of plants in different ecological zones. To evaluate the regions having potential for the growth and development of dragon fruit, research based on the field examinations, using the Linear Regression with 2 variables: elevation and temperature to find the relationship between temperature and elevation, thereby construct the map of temperature for the target area. With 10 data points observed in fieldtrip, the research obtained.

Equation of variability of temperature depending upon elevation in the winter:

$$y = -0.006x + 18,769$$

Equation of variability of temperature depending upon elevation in the summer:

$$y = -0.0054x + 32.517$$

where: y is temperature (°C); x is elevation (m).

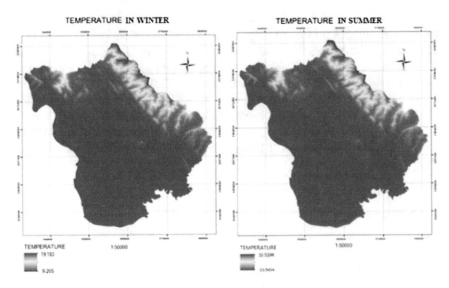

Fig. 2 Map of temperature in summer and winter

By the result of study about relationship between temperature and elevation, using DEM for Vinh phuc province, the research created temperature map for this area in the winter and summer (Fig. 2).

Precipitation
The research used the data of rain capacity monitored from monitor stations on the locality of Vinh Phuc province, as well as international monitoring data shared by Globalweather (NCEP). The precipitation map of the study area was interpolated by

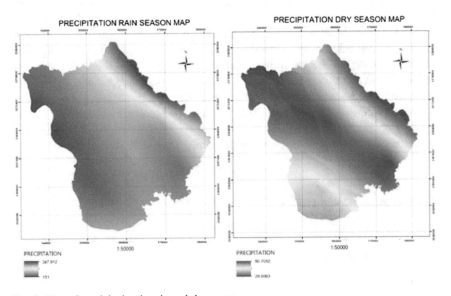

Fig. 3 Map of precipitation in rain and dry season

Ordinary Kriging from the average monitoring data of two seasons: rainy season (from April to September) and dry season (from October to March) (Fig. 3).

Slope

Slope is also an important factor in assessing suitability for selection of crop. Sloping land is often subjected to erosion, which leads to land degradation, soil degradation, poor nutrient, structural failure, pH reduction, increase of levels of soil toxicants and it can make the soil die biologically. Most of sloping lands are degraded and sour, many areas were deserted due to loss of agricultural and forestry production ability. For dragon fruit, the lower the slope, the more suitable for cultivation. In this study, the slope map was created based on the DEM of Vinh Phuc (Fig. 4).

Distance to Irritation Source, Distance to the Road

The distance to the irritation source is also taken into account in determining the suitability of cultivation. Based on the map of Red river in Vinh Phuc province, the study used the IDRISI tool to calculate the raster map showing the distance to

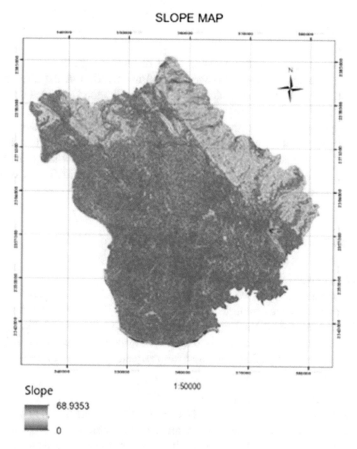

SLOPE MAP

Slope
68.9353
0

1:50000

Fig. 4 Map of slope

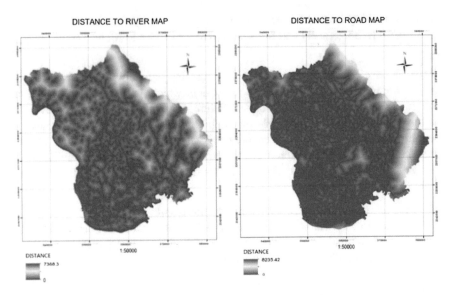

Fig. 5 Distance to the road and distance to irritation source

the irrigation source for each location. The distance to the irritation source for growing the dragon fruit should range from 20 m to 100 m. If the distance is less than 10 m, the location is not suitable for growing dragon fruit because it can be influenced by flooding and landslide. A distance of more than 100 m is also considered inappropriate due to the difficulty of water supply for irrigation.

For the purpose of harvesting, transporting products, the distance to roads is also included as an indicator to assess the suitability of dragon fruit cultivation. Research used optimal distance to the road from 20 to 100 m, over 5000 m is considered inappropriate (Fig. 5).

2.2 Evaluate, Synthesize and Analyze Data

Multiple-criteria decision analysis (MCDA)

Multiple-criteria decision analysis (MCDA) is a sub-discipline of operative research that explicitly evaluates multiple conflicting criteria in decision making. MCDA methods can be divided into 2 types: (i) multi-objective decision making (MODM) and (ii) multi-attribute utility theory (MAUT). This stage includes standardization, expert work, weighting calculation, and summary analysis of all criteria considered in the decision-making process

Standardize the evaluation factors

Since the datas collected are measured on different scales, the first step of MCDA is to standardize all data sets and comparable units. There are a large number of approaches that can be used to create attribute classes of comparable criteria. Based on the researches and experiences of the experts, in this study, the fuzzy set theory was used to standardize the criteria of the data. The input data is converted to the same scale from 0 to 255. The values of the standard data are determined by J-shaped and Linear functions (Table 1).

Analytic Hierarchy Process (AHP)

AHP is a decision-making support technique that provides an overview of the order of the design choices and help to find out the most appropriate final decision. The AHP helps decision-makers find what is best for them and helps them to understand their problems. Based on mathematics and psychology, AHP was developed by Saaty in 1970 and has been expanded and added to date. AHP provides a precise framework for structuring of a problem to be solved. AHP is tightly combined with decision standards and decision makers use pairwise comparison to determine reciprocally between goals. The AHP method is to construct the weight of the selected criteria for assessing adaptation of the red dragon fruit. From that, put the weights into the AHP to calculate and establish the appropriate partition map (Table 2).

Therefore, the consistency ratio CR = 0:06 satisfied the criterias, the average weight is confirmed and put into the index calculations incorporate the adaptive building zonation map (Table 3).

Table 1 Fuzzy standardize the evaluation elements

Factors	Function used	Control points
Rainfall in rainy season	J-Shape increasing	a = 50 mm, b = 150 m, c = 176 m, d = 270 m
Precipitation in dry season	J-Shape symmetric	a = 29 mm, b = 51 mm
Temperature in hot season	J-Shape increasing	a = 15 °C, b = 21 °C, c = 26 °C, d = 35 °C
Temperature in cold season	J-Shape symmetric	a = 15 °C, b = 19 °C
Slope	Linear decreasing	a = 0%, d = 17%
Distance to roads	Linear symmetric	a = 5 m, b = 20 m, c = 100 m, d = 8000 m
Distance to rivers, lakes, ...	Linear symmetric	a = 5 m, b = 20 m, c = 100 m, d = 4000 m

Table 2 Weight of factors

Factors	Weight
Rainfall in rainy season	0.178
Precipitation in dry season	0.1116
Temperature in summer	0.0958
Temperature in winter	0.4689
Distance to rivers, lakes, ...	0.0859
Slope	0.0371
Distance to roads	0.0228

Table 3 Parameters of AHP

Factors	Value
Eigenvalues of Matrix (λ_{max})	7.475
Factors (n)	7
Consistency index (CI)	0.079
Random (RI)	1.32
Consistency Ratio (CR)	0.06

3   Results

Results of the research are illustrated in the map of factors affecting dragon fruit in Vinh Phuc province, together with additional research methods to provide a suitable zonation map of each sites for dragon fruit planting (Fig. 6).

According to research results, it can be seen that most of Vinh Phuc province's areas have suitable conditions for red dragon fruit, as follows: highly suitable area covers an area of 30,628.17 ha, located in Song Lo, Lap Thach, Vinh Tuong, Phuc Yen districts; the area which is lower suitable for growing dragon fruit is of 1586.08 ha, located in the Song Lo, Lap Thach, Tam Dao, Phuc Yen districts. There are 62148.28 ha, which is completely not suitable for dragon fruit growth (Fig. 7).

The research explored the potential development of red dragon fruit in the management of resources in Vinh Phuc province. The results have contributed to improvements in methodology, time and cost for planning and resources

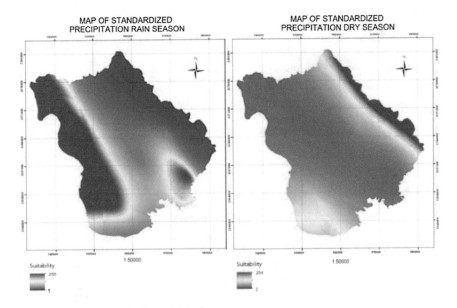

Fig. 6 Standardization map for precipitation

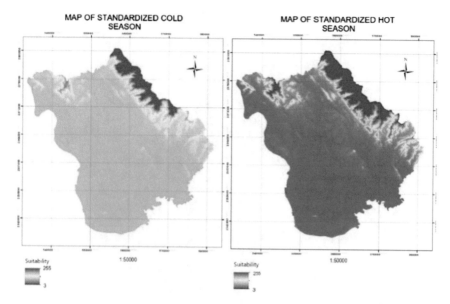

Fig. 7 Standardization map for temperature

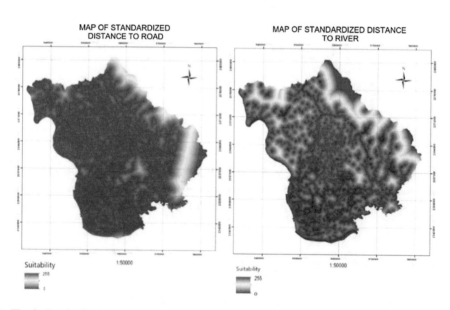

Fig. 8 Standardization map of distance to the road and to irritation source

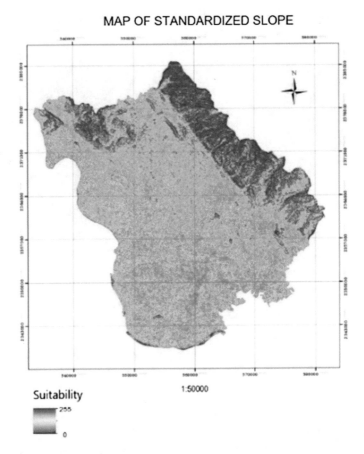

Fig. 9 Standardization map for slope

management. However, due to the difficulties of data collecting as well as the limitations of time and funding, the research only focused on a number of natural and social factors that affect growth of dragon fruit (Fig. 8).

4 Conclusion and Recommendation

To conclude, by using GIS and others analysis methods, the research showed that most of Vinh Phuc province's area are suitable for red dragon fruit growth (Fig. 9).

Based on the research results and the plan for agriculture, forestry and fishery development in Vinh Phuc province up to 2020 with a vision to 2030, the red dragon fruit is encouraged to develop for transforming forest land into economic

POTENTIAL MAP

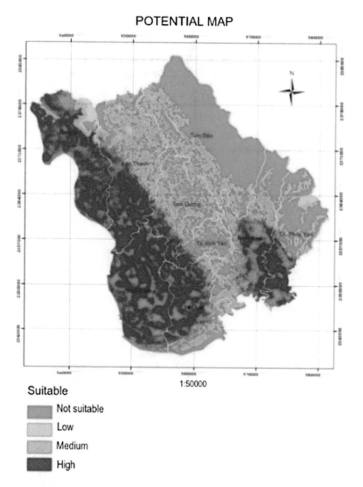

1:50000

Suitable

Not suitable

Low

Medium

High

Fig. 10 Map of potential development of red dragon fruit in Vinh Phuc province

models. Song Lo, Lap Thach, Phuc Yen district should be focused on growing this fruit plant (Fig. 10).

The methodology of the research not only use for assessment and zonation of plant development, but also can be applied in many different fields such as: warning of natural disasters, nature conservation and evaluation, planning and rational use of land resources.

References

Dep TT (2000) Identification of growth factors and long-lasting photogallery products for flower production for Dragon Fruit. PhD thesis, University of Natural Sciences, Ho Chi Minh City

Liaotrakoon W (2013) Characterization of dragon fruit (Hylocereus spp.) components with valorization potential. PhD thesis, Ghent University, Belgium, pp 6–20
Van Ke N (1998) Dragon fruit tree. University of Agriculture and Forestry, Ho Chi Minh City
Vinh Phuc Provincial People's Committee (2011) Vinh Phuc Province's agricultural, forestry and fishery development plan up to 2020 with a vision to 2030, Vinh Phuc

Evacuation Exercises and Simulations Toward Improving Safety at Public Buildings

Angela Santos, Margarida Queirós and Gabriele Montecchiari

Abstract An evacuation exercise was conducted at a teaching and research institute of Lisbon University, where 64 first year graduate students participated. Videos were recorded during the exercise in order to analyze the students' behavior. At the end of the activity students answered a questionnaire. In addition, computer simulations of the evacuation were carried out to compare the theoretical behavior with the experimental one. The results show that although about 90% of the students have participated in drills, some had an inappropriate behavior during the evacuation, such as stopping or walking too slowly. Simulation results show that these kinds of behavior delayed the evacuation by about 5 s. Furthermore, the comparison between experimental results and simulations has highlighted the necessity of studying and developing models capable of reproducing, to some extent, those behaviors considered to be incorrect during an evacuation. These kinds of models might be used to assess the evacuation time considering also the level of training of the population.

Keywords Evacuation exercises · Videos · Simulations · Emergency plan

A. Santos (✉) · M. Queirós
Centre for Geographical Studies, Institute of Geography and Spatial Planning, Universidade de Lisboa, Rua Branca Edmée Marques, Edifício IGOT, 1600-276 Lisbon, Portugal
e-mail: angela.santos@campus.ul.pt

M. Queirós
e-mail: margaridav@campus.ul.pt

G. Montecchiari
Department of Engineering and Architecture, University of Trieste, Building, C5 – Floor II Via A. Valerio 10, 34127 Trieste, Italy
e-mail: gabriele.montecchiari@phd.units.it

© Springer International Publishing AG 2018
B. Otjacques et al. (eds.), *From Science to Society*, Progress in IS,
DOI 10.1007/978-3-319-65687-8_3

1 Introduction

Previous studies have shown the importance of the implementation of regular practice of evacuation exercises at educational institutions (Santos and Queirós 2017; Gu et al. 2016), especially related to earthquakes and fires. Although there are many national and international guidelines related to evacuation exercises and drills, as compiled by (Santos and Queirós 2017), literature related to the evacuation behavior is limited. Thus, this study was carried out under the guidelines provided by the Portuguese National Authority of Civil Protection. Such guidelines recommend performing earthquake evacuation exercises, annually on November 5th (http://www.aterratreme.pt). Thus, this research is a contribution to practice: by following these guidelines, the authors hope to promote the integration of students in disaster management activities with regard to communication of the earthquake and fire risks. The aim is to build disaster resilient educational communities, by improving the capabilities of individual students to respond to an emergency situation. Therefore, the objectives of this study are to present and discuss the results of this evacuation exercise. With this research the authors would like to increase awareness related to this subject as well as disseminate the outcomes for future activities.

2 Data Collection

During the first week of November 2016 an experiment was conducted involving first year graduate students at a teaching and research institute of Lisbon University (IGOT). The students were distributed over three classes, thus the experiment was conducted on three independent times. The methodology adopted for this activity followed previous studies (Santos and Queirós 2017): first, students were briefed about basic information related to earthquakes and tsunamis and safety procedures at the university campus, which included the indication of evacuation routes and the location emergency signs and equipment. Then, they carried out the evacuation exercise; the trigger of the evacuation was a hypothetic earthquake. The students followed the safety instructions: protected themselves under the tables, and then they walked calmly till the meeting point (assembly point), located outside of the building. Finally, at the meeting point students answered a questionnaire. Furthermore, videos were recorded during the evacuation, which started inside the classroom and ended at the meeting point (assembly point), located outside the building. This approach has been used in order to understand the behavior of evacuees during the exercises (Gu et al. 2016).

3 Simulation Methods

As a complement to data collection, simulations were also carried out for comparison with experimental data. Detailed measurements of the room were collected, which included the dimensions, width of the exit door, as well as position and dimensions of the desks. These data were provided as input to the evacuation simulator developed in (Montecchiari et al. 2016; Montecchiari et al. 2017) which has been developed targeting the maritime field, but can be adopted also in the field of buildings evacuation. The software is based on the social force model developed by Helbing and Johansson (2011) and on its FDS+Evac implementation as described by Hostikka (2009). The model is agent based, meaning that each evacuee is modeled separately with its own properties. The agents path, in absence of interactions, is determined through the use of waypoints and of a shortest path algorithm based on the implementation presented in (Granberg 2017). Agents properties govern their interaction with the environment and with other agents. In this specific case the agents' properties and initial conditions have been set in accordance with the data acquired during the experiments. The distribution of the unimpeded speeds of the agents has been modeled as a uniform distribution. The intervals for the unimpeded speed of the modeled agents have been defined starting from those provided by the International Maritime Organization MSC.1/Circ.1533 (IMO 2016) for 'Males younger than 30' and 'Females younger than 30', and rescaled according to the average unimpeded speed of 1.06 m/s which was measured during the present experiments. The agents' mass has been set according to the data acquired during the experiment, whereas the agents' dimensions have been obtained by rescaling the default case reported by Hostikka (2009) according to the measured mass. The modeling in the adopted tool embeds a certain level of aleatory uncertainty; therefore there is the need to acquire a sufficient number of realizations to obtain a reliable statistics. To this purpose, 100 Monte Carlo simulations have been performed where the unimpeded speed varies according to the agents' sex and initial position, while the mass and dimensions have been left unchanged.

4 Experimental Results and Discussion

A total of 64 students participated in the activity, distributed within 3 classes. However, there were 61 replies to the questionnaire, and the results are presented in Tables 1, 2 and 3. The first group (Table 1) indicates that most students are males (67.2%) and the most common ages are 19–20 years old (55.7%). The weight and height were asked to provide input data to the simulations. The motivation of the students to participate in the activity is lower than expected since about 46% indicated "very high" or "high. In addition, more than 90% of the students are healthy, have participated in evacuation exercises before, have never been in a situation of risk and know the emergency number. However, only about 10% have

Table 1 Questionnaire results of the first group of questions (Personal Information)

Question	No.	%	Question	No.	%
1.1 Gender			**1.6 Sickness that affects safe evacuation**		
Male	41	67.2	Yes	4	6.6
Female	20	32.8	No	56	91.8
1.2 Age			No answer	1	1.6
17–18	18	29.5	**1.7 Mobility problem**		
19–20	34	55.7	No	61	100.0
21 or more	9	14.8	1.8 Participation in a drill:		
1.3 Weight (kg)			Yes	55	90.2
45–60	24	39.3	No	6	9.8
61–70	21	34.4	**1.9 Use of fire extinguisher**		
71–80	10	16.4	Yes	6	9.8
81 or more	6	9.8	No	55	90.2
1.4 Height (m)			**1.10 Ever felt any risk**		
1.50–1.60	9	14.8	Yes	4	6.6
1.61–1.70	17	27.9	No	55	90.2
1.71–1.80	24	39.3	No answer	2	3.3
1.81 or more	11	18.0	**1.11 Emergency phone numbers (several options)**		
1.5 Motivation			Emergency	55	90.8
Very high	8	13.1	Local Firefighters	0	0
High	20	32.8	Local Police	0	0
Sufficient	26	42.6	IGOT security	0	0
Low/No motivation	7	11.5	No answer	6	9.8

Table 2 Questionnaire results of the second group of questions (Drill)

Question	No.	%	Question	No.	%
2.1 Followed the initial route			**2.3 Indication of fire extinguishers**		
Yes	58	95.1	0	21	34.4
No/No answer	3	4.9	1–10	37	60.7
2.2 Indication of emergency signs			11 or more	3	4.9
0	17	27.9	**2.4 Indication of emergency buttons**		
1–10	8	13.1	0	17	27.9
11–20	23	37.7	1	4	6.6
21 or more	13	21.3	2–3	40	65.6

used a fire extinguisher. Still, less than 10% of the students did not answer questions 1.10 and 1.11, which is consistent with the lack of attention and motivation (question 1.5).

Table 3 Questionnaire results of the third group of questions (Evaluation)

Question	No.	%	Question	No.	%
3.1 Feel safer after this exercise			**3.5 Repetition of the exercise**		
Yes	57	93.4	Yes	37	60.7
No	4	6.6	No	24	39.3
3.2 More prepared to act in emergency			**3.6 Situations to practice (several options)**		
Yes	59	96.7	Flood	29	47.5
No	2	3.3	Fire	26	42.6
3.3 Initial emergency plan adequate			Gas leak	23	37.7
Yes	59	96.7	Earthquake	14	23.0
No answer	2	3.3	No future practice	9	14.8
3.4 recommendation to others			No answer	1	1.6
Yes	60	98.4			
No	1	1.6			

Regarding the second group of questions (Table 2), all the students followed the initial route, although 4.9% said they did not or they did not provide any answer. In addition, more than 65% of the students were aware of the emergency signs and equipment. These results show that some students were distracted which is in agreement with the lack of motivation detected in the first group presented in Table 1. Furthermore, even motivated students had some difficulty in memorizing the location of the 28 emergency signs (question 2.2), 14 fire extinguishers (question 2.3), and 3 emergency buttons (question 2.4). Although a small number of students participated in the experiment, the results can be generalized, as discussed previously (Santos and Queirós 2015). In fact, a recent research shows that although students have been receiving information about safety procedures, they have low awareness about this topic (Santos et al. 2017). Thus more efficient ways to explore mental maps should be attempted. For this reason, more educational activities regarding safety must be conducted.

In the evaluation of the activity (Table 3), almost all the students (more than 93%) feel safer in the building, more prepared to act in emergencies, think the emergency plan is adequate and recommend it to others. They also would like to repeat the exercise (60.7%) and would like to practice the occurrence of floods (47.5%), fires (42.6%) and gas leaks (37.7%). From the analysis of the answers to this group of questions it is also clear that about 16% of the students were not motivated to participate in future activities. This outcome is consistent to the previous group questions.

On the other hand, video analysis of classes 1, 2 and 3 showed interesting behavior of the students. For example, although about 90% of the students have participated in drills (question 1.8), some had a unexpected behavior during the evacuation: in class 1 some males stopped to allow females to pass the exit door

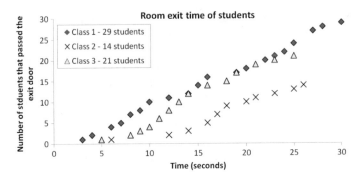

Fig. 1 Summary of experimental data

(cultural behavior), while other students simply stopped for no reason, delaying the evacuation; in class 3, a female stopped in front of the door, indicating that she was waiting for her friends; in all classes, the first student left the room quickly, between 3 and 5 s after the evacuation started (Fig. 1) but in class 2 they were slower than the ones in classes 1 and 3. Also in all classes, students left the classroom in groups of 2 or 3. Nevertheless, the evacuation of the room was conducted in less than 30 s, and the total evacuation time till the meeting point located outside of the building was about 1 min and 30 s.

5 Simulation Results and Discussion

The agents' initial positions of class 2 and the geometry used for the simulation are reported in Fig. 2. Figure 3 reports snapshots of one simulation for a qualitative comparison with experimental outcomes. From the simulation we observe a continuous flow towards the room exit, which is only slowed down by the bottleneck condition occurring at the door. However, from the video, it is clearly visible that the flow of students of class 2 is often interrupted. These interruptions are either due to the presence of chairs in the room or by the fact that some students occasionally stopped, thereby blocking the flow. This behavior is confirmed by the quantitative observation of exit times in Fig. 4a, which reports the exit time of each agent and compares them with the experimental exit time of the corresponding evacuee.

The exit times resulting from the simulations match the experimental ones, for more than half of the evacuees. In the other cases, instead, the evacuees exit time in the experiment is much longer than the one obtained from simulation. An explanation for the observed discrepancies can be given by looking at the recorded video. In fact, the video shows that evacuee 2 stops before the exit in order to wait for evacuees 3 and 4. Evacuees 3 and 4 were left behind because they stopped to take their own bags. Those kinds of behaviors might affect the overall time performance

Fig. 2 Initial positions of evacuees in the experiment with Class 2 and in simulations

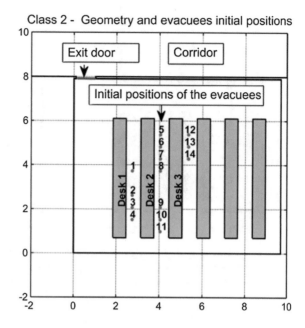

Class 2 - Geometry and evacuees initial positions

of the evacuation process, and they are not part of the modeling used in the simulations.

However, the phenomenon where, due to geometrical constraints, one evacuee slows down all the other can be observed also in the simulations outcomes. To this purpose those simulation where the evacuees 1, 5 and 12 obtain the largest exit time were highlighted in black, red and green respectively in Fig. 4a. The figure shows that in each desks line the exit time of the evacuees is affected by the reduced speed of the evacuee occupying the initial position of the line, as it was expectable. The effect, on the overall exiting performance, of the lack of complete modeling of evacuees' behaviors can be better appreciated looking at the exit time curve reported in Fig. 4b. Apart from an initial delay, it can be observed that the flow rate of the evacuees in the experiment matches well the one obtained from the simulations.

As anticipated, it can be noted that, early in the evacuation process, the experimental curve is shifted of some seconds with respect to the median of simulations. Instead, the trend of the subsequent part of the experimental exit rate is very similar to that obtained from simulations. The initial shift is due to the initial slowdown of the evacuation process, which is present in the experiment but not in any of the simulations. This shift corresponds, in the video, to the situation of an evacuee stopping in front the exit while waiting for two others evacuees.

Behaviors such as stopping for apparently no reason or going in the opposite direction with respect to the exit are generally not considered when the assessment of evacuation time is performed through the use of simulators. For instance, there is no requirement by IMO concerning the modeling of these behaviors (IMO 2016). This is mainly due to the fact that models simulating these behaviors are very

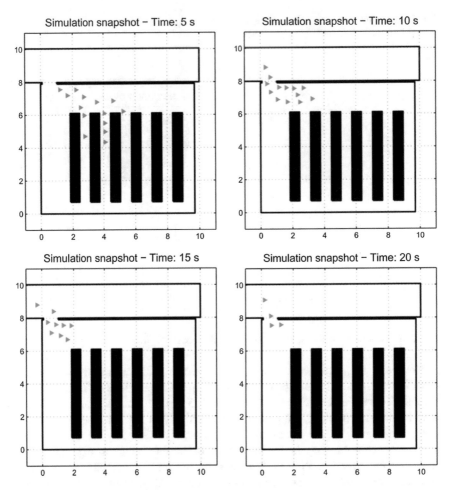

Fig. 3 Snapshot from one simulation (class 2)

difficult to calibrate and validate. Moreover, such unpredictable behaviors might change drastically if considering a different population or different conditions (for example a real condition of danger).

However, the comparison between theoretical and experimental outcomes highlighted the possibility of easily detecting incorrect behaviors and their effects on the overall evacuation process. Moreover the potential of the adopted simulation tool have not been yet fully explored. The tool can indeed be used also to model larger panels of people and more complex geometries (e.g. staircases, multiple floors), still allowing to obtain reasonably realistic behaviors, even though a proper validation in complex cases is difficult to carry out also due to the lack of detailed experimental results in literature. Comparisons with similar simulation tools, thereby, might be useful to assess the evacuation skills of the subjects involved in the evacuation drill in terms of evacuation best practices and improve subjects'

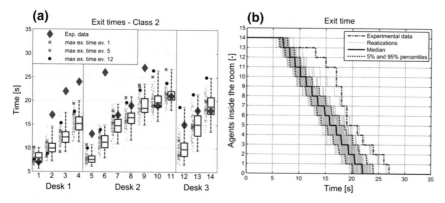

Fig. 4 a Exit times resulting from 100 realizations. Results are reported for each agent through box plots (min, 25%, 50%, 75%, max), scatter data, and compared with the corresponding experimental outcome. Some simulations are highlighted. In particular are highlighted the simulation outcomes associated with simulations having the maximum exit time for the evacuee 1 (*black*), the evacuee 5 (*red*) and the evacuee 12 (*green*). Evacuees 1, 5 and 12 are the ones positioned initially at the beginning of their respective desk line, thereby their performance, in terms of exit time, affects also the performance of all the agents in the same line; **b** Simulated exit time and comparison with experimental data. Outcomes from all realizations outcomes are reported together with ensemble median and 5% and 95% percentiles

performance by the identification of the incorrect behaviors. In fact, recent urban fires that have occurred in Portugal in a hotel (de Notícias 2017), hospital (Jornal Público 2017) and nearby a supermarket (RTP 2017) have shown that a proper evacuation plan is very effective in directing evacuees safely outside of a building. Therefore, although almost all students reported that the initial emergency plan was adequate (question 3.3), it must be improved, and more evacuations exercises must be carried out.

6   Conclusions

The results of the questionnaires show that 46% of the students were motivated to the activity, but in the end about 61% would like to repeat the exercise for floods, fires and gas leaks. In addition, almost all the students (more than 93%) feel safer in the building, more prepared to act in emergencies and recommend it to others. Still, about 16% did not answer questions, showed lack of attention, motivation and weren't interested in repeating this king of activity. Therefore, future activities involving first year graduation students should be prepared in the context of all IGOT's users, with involvement of civil protection agents, or more briefings on the subject. Furthermore, even motivated students had some difficulty in memorizing the location of emergency signs and equipment, thus more efficient ways to explore mental maps should be attempted and more exercises must be conducted.

Although more than 90% of the students have participated in previous evacuation exercises in other contexts, video analysis shows that some students had an inappropriate behavior, such as moving too slowly, stopping for no reason, or waiting for friends. These kinds of behaviors delay the evacuation. Furthermore, simulations show that the unprepared students' behavior actually delays the evacuation for about 5 s. In this simple experiment, it was not significant because the number of evacuee and the complexity of the geometry is relatively small. However, if the entire building population had to evacuate, chaos might occur.

For all the above reasons, more educational activities regarding safety should be conducted, including the regular practice of evacuation exercises and improvement of emergency plan. Furthermore, the comparison between experimental results and simulations has highlighted the necessity of studying and developing models capable of reproducing, to some extent, those behaviors considered to be incorrect during an evacuation. These kinds of models might be used to assess the evacuation time considering also the level of training of the population.

Acknowledgements The authors would like to thank the IGOT students that participated in the exercise, and the professors who allowed this experiment in their classrooms and even kindly recorded the videos. The authors also would like to thank Gabriele Bulian (University of Trieste) for the useful suggestions to the paper. The research activity of the third author has been carried out in the framework of his Ph.D. program at University of Trieste (Ph.D. scholarship G/3-Cycle XXX financed by MIUR "Progetto Giovani Ricercatori" through University of Trieste).

References

de Notícias JD (2017) Fire forces to evacuate Hotel Sheraton, 14 Feb 2017 (in Portuguese)

Granberg A (2017) http://arongranberg.com/astar/. Accessed 6 June 2017

Gu Z, Liu Z, Shiwakoti N, Yang M (2016) Video-based analysis of school students' emergency evacuation behavior in earthquakes. Int J Disaster Risk Reduct 18:1–11

Helbing D, Johansson A (2011) Pedestrian, crowd and evacuation dynamics. Extreme environmental events—complexity in forecasting and early warning. Springer, New York, pp 697–716

Korhonen T. Hostikka S (2009) Fire dynamics simulator with evacuation: FDS+Evac. Technical reference and user's guide. VTT Technical Research Centre of Finland, FDS+Evac

IMO (2016) MSC.1/Circ.1533—Revised guidelines for evacuation analysis for new and existing passenger ships, 6 June

Jornal Público (2017) Fire at the São Francisco Xavier Hospital forces 15 people to be assisted, 5 Apr 2017 (in Portuguese)

Montecchiari G, Bulian G, Gallina P (2016) Development of a new evacuation simulation tool targeting real-time human participation. In: Walls, Revie, Bedford (eds) Risk, reliability and safety: innovating theory and practice. CRC Press, Balkema, pp 571–578

Montecchiari G, Bulian G, Gallina P (2017) Towards real-time human participation in virtual evacuation through a validated simulation tool. J Risk Reliab (JRR) (to appear)

RTP (2017) Small aircraft falls nearby a supermarket, 17 Apr 2017 (in Portuguese)

Santos A, Queirós M (2015) Public buildings safety: addressing a pilot evacuation exercise. In: Nowakowski et al (eds) Safety and reliability: methodology and application. Taylor & Francis Group, London, pp 2009–2015. ISBN 978-1-138-02681-0

Santos A, Queirós M (2017) Safety procedures at university campus: the implementation of evacuation exercises. In: Walls, Revie, Bedford (eds) Risk, reliability and safety: innovating theory and practice. Taylor & Francis Group, London, pp 2268–2273. ISBN 978-1-138-02997-2

Santos A, Queirós M, Carvalho L (2017) Fire and seismic risk perception at Lisbon University—Faculty of Letters. Territorium 24:15–27

Part II
Environmental Modelling and Simulation

Estimating the Environmental Impact of Agriculture by Means of Geospatial and Big Data Analysis: The Case of Catalonia

Andreas Kamilaris, Anton Assumpcio, August Bonmati Blasi, Marta Torrellas and Francesc X. Prenafeta-Boldú

Abstract Intensive farming has been linked to significant degradation of land, water and air. A common body of knowledge is needed, to allow an effective monitoring of cropping systems, fertilization and water demands, and impacts of climate change, with a focus on sustainability and protection of the physical environment. In this paper, we describe AgriBigCAT, an online software platform that uses geophysical information from various diverse sources, employing geospatial and big data analysis, together with web technologies, in order to estimate the impact of the agricultural sector on the environment, considering land, water, biodiversity and natural areas requiring protection, such as forests and wetlands. This platform can assist both the farmers' decision-taking processes and the administration planning and policy making, with the ultimate objective of meeting the challenge of increasing food production at a lower environmental impact.

Keywords Policy tool · Agriculture · Environmental impact · Geospatial analysis · Big data analysis

A. Kamilaris (✉) · A. Assumpcio · A.B. Blasi · M. Torrellas · F.X. Prenafeta-Boldú
GIRO Joint Research Unit IRTA-UPC, Barcelona, Spain
e-mail: andreas.kamilaris@irta.cat; camel9@gmail.com

A. Assumpcio
e-mail: Assumpcio.Anton@irta.cat

A.B. Blasi
e-mail: august.bonmati@irta.cat

M. Torrellas
e-mail: marta.torrellas@irta.cat

F.X. Prenafeta-Boldú
e-mail: francesc.prenafeta@irta.cat

© Springer International Publishing AG 2018
B. Otjacques et al. (eds.), *From Science to Society*, Progress in IS,
DOI 10.1007/978-3-319-65687-8_4

1   Introduction

The central role of the agricultural sector is to provide adequate and good-quality food to an increasing human population and, because of its importance and relevance, it is on the focus of the global policy agendas.

Agriculture is considered an important contributor to the deterioration of soil, water contamination as well as air pollution (Bruinsma 2003). Intensive farming has been linked to excessive accumulation of soil contaminants (Teira-Esmatges and Flotats 2003), as well as to significant groundwater pollution with nitrate (Stoate et al. 2009). Almost 23% of global greenhouse gas emissions are attributed to agriculture. The negative environmental impact from livestock farming across Europe continue to make their mark (Heinrich Böll Stiftung 2014), resulting in new legislations, policies and large research programs. However, despite a huge amount of published material and many available techniques, doubts over the success of national and European initiatives remain, regarding the extent over which environmental targets are met (Loyon et al. 2016). Hence, a common body of knowledge is necessary to be developed, shared at local and regional levels of countries involved and affected, so as to allow an effective monitoring of cropping and animal production systems, fertilization and water demands, and impacts of climate change. This knowledge would assist policymakers to perform efficient regulatory enforcement considering sustainability and protection of the physical environment. To acquire this knowledge, web applications and mobile apps need to be combined with geospatial and big data analysis (Mintert et al. 2016), in order to develop online tools that allow policymakers to perceive, visualize and analyze the impact of agriculture, facilitating decision making towards mitigating or eliminating negative effects on the environment. Big data analysis is crucial for analyzing vast amounts of data (e.g. weather, air and water quality, pests and animal diseases etc.) coming from various sources in near real-time. Geospatial analysis is also important for large-scale planning while web/mobile apps can provide access to the users by means of graphical, user-friendly interfaces. In this paper, we describe how the aforementioned technologies (geospatial and big data analysis, web/mobile apps) can be combined to develop an online tool for policymakers (and possibly to farmers in the future as well), towards estimating the impact of agriculture on the environment. This is one of the first initiatives (in the agri domain) combining these technologies together in order to address such a complex multivariate problem. Hence, the contribution of this work is to describe this policy tool (Sect. 3), focusing on the area of Catalonia, Spain during our analysis efforts (Sect. 4).

2 Related Work

The majority of software in the agri-domain involves modeling software and simulations for soil, crops, water needs, adaptation to climatic change etc., targeting mostly the farm level (Holzworth et al. 2015). More related efforts include geospatial platforms and applications focusing on various problems of the agricultural sector, such as facilities management (Lucas and Chhajed 2004), spatial decision support systems (Silva et al. 2014), land use and land cover changes (Embrapa 2016), diseases control and epidemiology (Cringoli et al. 2007) etc. Most relevant work involves geospatial platforms dealing particularly with the impact of agriculture on the environment. Examples include animal manure transportation (Paudel et al. 2009), selecting sites for safe application of animal waste as fertilizer to agricultural land (Basnet et al. 2001) and environmentally sound management of livestock production (Jain et al. 1995). Finally, big data management and analysis platforms have recently appeared, focusing either on climate change (Schnase et al. 2017) or earth observation (Nativi et al. 2015).

3 AgriBigCAT Platform Description

AgriBigCAT is an online software platform combining geophysical information from various diverse sources and web technologies, together with geospatial and big data analysis, in order to estimate the impact of the agricultural sector on the environment, considering land, water, air emissions (e.g. greenhouse gases), biodiversity and natural areas such as forests and wetlands. AgriBigCAT intends to promote more sustainable agriculture, constituting a knowledge-based platform, managing and analyzing a wide range of geospatial and sensory/multimedia information, accessible by standard communication technologies such as the internet/web and mobile apps. Its architecture is depicted in Fig. 1.

The platform allows large-scale data acquisition and analysis of relevant parameters in various agricultural systems in near real-time, facilitating geospatial management and use of inputs (e.g. energy, nutrients and water) and outputs (e.g. emissions, biomass yield, etc.). AgriBigCAT assists both farmers' decision-making and administration planning/policy making, with the objective of meeting the challenge of increasing food production at a lower environmental impact. Agriculture-related datasets are stored using Apache Hive, a (big) database software that allows management of large datasets residing in distributed storage. Visualizations of the datasets as well as geospatial analysis are performed by means of ArcGIS (Esri 2017) and its API for JavaScript, which allows AgriBigCAT to be developed by using open web technologies (e.g. HTML, CSS, JavaScript, AJAX, PHP), but at the same time use the visualization and geospatial features of ArcGIS through its API.

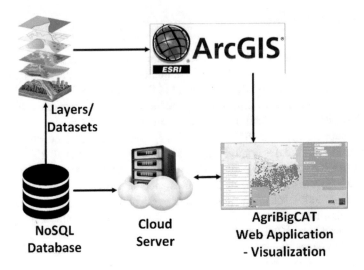

Fig. 1 AgriBigCAT architecture

Table 1 AgriBigCAT operations, technologies and big data characteristics

No.	Operation	Technology	Big data characteristic
1.	Data storage	Apache Hive	Volume (V1), Variety (V3)
2.	Data analysis	ArcGIS Cloud with Hadoop	Volume (V1), Velocity (V2), Variety (V3)
3.	Data visualization	AgriBigCAT	Velocity (V2), Valorization (V5)

Finally, specific, "heavier" scenarios of geospatial analysis are executed by means of ArcGIS and illustrated through AgriBigCAT in the form of *maps*. In this case, datasets imported into ArcGIS are abstracted as *layers*. Over time, multiple layers become part of a large-scale geo-database for spatiotemporal analysis. Table 1 lists the operations and technologies used by AgriBigCAT, according to big data characteristics as defined in (Chi et al. 2016). Data sources include geospatial datasets ($\sim$GB), sensor measurements ($\sim$MB), historical data in various file formats ($\sim$MB) and images ($\sim$GB–TB). Data veracity (V4) is assured by collaborating with reputable agencies for data collection, such as the Ministry of Agriculture of Catalonia (see Sect. 4).

4 Analysis

We have primarily focused on the territorial domain of Catalonia (Spain) motivated by the fact that it is one of the European regions with the highest livestock density, with farm concentrations of more than 6 M pigs, 0.7 M cows and 38 M poultry. The high density of livestock in some areas, linked to the insufficient accessible arable land, has resulted in severe groundwater pollution with nitrates (Nitrates Directive 1991). Excessive soil accumulation of phosphorous and heavy metals from manure has also been reported in certain areas (Teira-Esmatges and Flotats 2003). The main purpose of this analysis is to optimize manure management by proper planning, aiming to prevent manure land overdosing eliminating negative effects on the environment. The analysis could also result in optimization on the application of agricultural inputs by means of more efficient policies and guidance. As a first step, large and diverse information datasets of the territorial domain of Catalonia were collected, summarized in Table 2. Difficulties in data collection are discussed in (Kamilaris 2017).

As described in Sect. 3, these datasets were imported as layers in ArcGIS for geospatial analysis, as well as in our NoSQL database for real-time calculations/estimations. Combining these two aforementioned features/services with web technologies, we developed a demo application, available online (P-Sphere Project 2017), allowing for policymakers, technical advisory services, researchers and environmental scientists in Catalonia to estimate the impact of animal production on the physical environment. Specifically, users can select a particular area on Catalonia (or all areas), particular animal type (e.g. pigs, dairy cows, poultry, beef cattle or all animals), and emission type (e.g. carbon dioxide,

Table 2 Datasets collected for the application scenario

No.	Area/Group	Datasets
1.	People, farms and animals	Manure management units, Farmers of Catalonia and numbers/types of animals they possess, Habitats of Catalonia
2.	Areas and land	Fishing, Evapotranspiration/thermal regions, Nitrate vulnerable areas, Forests, Municipalities, Land parcels, Soils, Crops
3.	Infrastructures	Wind parks, Road networks, Water/gas pipeline network
4.	Biodiversity	Vertebrates, Lormophytes, Lichens, Bryophytes, Waterbirds
5.	Climate and atmosphere	Thermal levels, Temperature, Rainfall, Noise maps, Climate type, Atmospheric emissions (CO_2)
6.	Water	Wetlands, Zones vulnerable in nitrates, Wastewater plants, Watersheds, Rivers, Reservoirs, Water network, Monitoring stations, Lakes, Water-deficit areas, Coasts, Bays, Aquifers

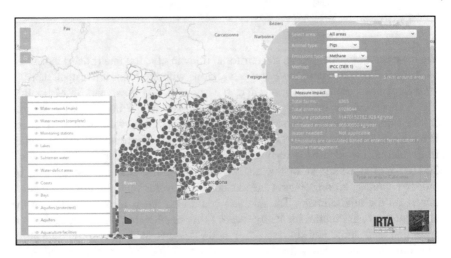

Fig. 2 Snapshot of the demo application

methane, nitrous oxide, ammonia) or nitrogen/phosphorous excreted, and the platform calculates the farms and animals involved in the query, manure produced and estimated emissions/excretions in monthly or yearly basis, taking into account the existing weather and thermal conditions at the selected time period from historical data. Calculations are based on the (IPCC 2006) guidelines (TIER1) while data about farms and animals have been provided by the Ministry of Agriculture of Catalonia (for the year 2016). Calculations also consider animal feeding practices widely used in the area, as well as popular manure management techniques employed in Catalonia. All assumptions involved are listed on the website, next to each visualization.

Figure 2 illustrates a snapshot of the demo application, where farm locations and rivers of Catalonia are visualized, in order to examine the impact of pigs' manure in relation to possible contamination of nearby rivers with nitrates.

Table 3 lists some visualizations and explanations of the geospatial problems addressed through our application. Our findings are being continuously reported to the Ministry of Agriculture of Catalonia for further actions.

Specific details over each of the above geospatial scenarios (e.g. methods, analysis, inputs, parameters, results) are available on the project's website (P-Sphere Project 2017). Some "sensitive information" is currently available only to authorized users, such as employees of the Department of Agriculture.

Table 3 Geospatial scenarios in Catalonia and visualizations/solutions

	Visualization of farms in Catalonia according to the type of animals they grow (i.e. pigs, dairy cows, cattle beef, poultry etc.). In this particular scenario, pig farms are displayed, which constitute the most popular farming industry in the region.
	Farms per municipality according to number of animals. Darker green means a larger concentration of farms/animals. Same maps have been created per animal type as well. The south-west part of the region has the highest concentrations.
	Hot spots of farms according to the total number of animals they have. It provides an indication of where high/low numbers of animals cluster spatially. Same maps have been created per particular animal type as well.
	Summary of yearly methane emissions (TIER2) at each municipality, based on existing farms/animals. Darker orange means a larger production of methane emissions, while the darkest one indicates that caution is required. Similar maps have been produced for other emissions such as nitrous oxide and ammonia. Same maps have been created per particular animal type as well. Calculations are based on the existing manure management treatment methods in the region.
	Summary of yearly nitrogen excreted from animals' manure at each municipality of Catalonia, based on existing farms and animal types. Similar maps have been produced for phosphorous too. Darker blue means a larger amount of nitrogen excreted. Same maps have been created per particular animal type as well. Calculations are based on IPCC TIER1. Similar maps will be created for the potential reduction of nitrogen/phosphorous by applying best practices (future work).

	Which are the best routes to connect farms together, e.g. for finding the best transportation solution to collect and carry manure by means of trucks? The best routes (i.e. minimizing total travel time/distance) are displayed in light brown color for the case of poultry farms. Similar maps have been created for other animal types and for all farms/animal types.
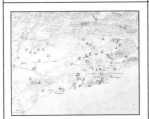	According to the structure of the existing manure management plants, which are the closest plants that can serve the animal farms in Catalonia to dispose their manure? The best plans are displayed as yellow rectangles for the case of poultry farms. Similar maps have been created for other animal types and for all farms/animal types.
	If we selected specific municipalities/locations to build manure processing units, considering their proximity to farms, where would these be? The best options (i.e. minimizing total travel time/distance) for the case of poultry farms are displayed as blue circles on the map. Similar maps exist for other animal types and for all farms.
	Which are the possibilities of using manure as crops' fertilizers? Is there an actual need per municipality according to crops produced and animals' manure being produced at the same time? The picture shows a matching between livestock farms and nearby crop-based farms that could use manure as fertilizer. (Work in progress)
	In which degree are areas vulnerable of nitrates affected by livestock farms? In which degree are rivers and lakes (water vulnerable zones) affected? The picture shows livestock farms and their proximity/impact to nearby rivers/lakes and nitrate vulnerable areas. (Work in progress)

5 Conclusion

This paper has described AgriBigCAT, an online software platform combining geospatial and big data analysis, together with web technologies, to estimate the impact of the agricultural sector on the environment. Serving as a knowledge-based platform, it constitutes a useful tool for administration planning and policy making, contributing to the challenge of increasing food production at a lower environmental impact. Moreover, an online application of AgriBigCAT, focusing on the local environmental issues of the agricultural sector of Catalonia, has been presented and described. As future work, we plan to examine best available techniques (BAT) available in agriculture for animal/manure management, and assess the potential environmental benefits (considering costs' trade-offs) of their application in Catalonia (or other regions considered in the future through our platform).

Acknowledgements This research has been supported by the P-SPHERE project, which has received funding from the European Union's Horizon 2020 research and innovation programme under the Marie Skodowska-Curie grant agreement No 665919.

References

Basnet BB, Apan AA, Raine SR (2001) Selecting suitable sites for animal waste application using a raster GIS. Environ Manage 28(4):519–531

Bruinsma J (2003) World agriculture: towards 2015/2030: an FAO perspective. s.l.: Earthscan

Chi M, Plaza A, Benediktsson JA (2016) Big data for remote sensing: challenges and opportunities. Proc IEEE 104(11):2207–2219

Cringoli G et al (2007) Geo-referencing livestock farms as tool for studying cystic echinococcosis epidemiology in cattle and water buffaloes from southern Italy. Geospatial Health 2(1):105–111

Embrapa (2016) SOMABrasil. http://mapas.cnpm.embrapa.br/somabrasil/webgis.html

Esri (2017) https://developers.arcgis.com/javascript/. http://www.esri.com/arcgis/about-arcgis

Heinrich Böll Stiftung (2014) Meat Atlas: Facts and figures about the animals we eat. Heinrich Böll Stiftung. Friends of the Earth Europe

Holzworth D et al (2015) Agricultural production systems modelling and software: current status and future prospects. Environ Model Softw 72:276–286

IPCC (2006) Chapter 10: Emissions from livestock and manure management. s.l.: IPCC Guidelines for National Greenhouse Gas Inventories

Jain DK, Tim US, Jolly RW (1995) A spatial decision support system for livestock production planning and environmental management. Appl Eng Agric 11(5):711–719

Kamilaris A (2017) Big data analysis and integration of geophysical information from the Catalan Agri-Technological sector. Interest Group on Agricultural Data (IGAD), Barcelona. https://www.slideshare.net/kamiWeb/big-data-analysis-and-integration-of-geophysical-information-from-the-catalan-agritechnological-sector

Loyon L et al (2016) Best available technology for European livestock farms: availability, effectiveness and uptake. J Environ Manage 166:1–11

Lucas MT, Chhajed D (2004) Applications of location analysis in agriculture: a survey. J Oper Res Soc 55(6):561–578

Mintert J et al (2016) The challenges of precision agriculture: is big data the answer? Southern Agricultural Economics Association Annual Meeting, San Antonio, Texas (No. 230057)

Nativi S et al (2015) Big data challenges in building the global earth observation system of systems. Environ Model Softw 68:1–26

Nitrates Directive (1991) Council Directive 91/676/EEC of 12 December 1991 concerning the protection of waters against pollution caused by nitrates from agricultural sources. EUR-Lex 375(31):1–8

Paudel K, Bhattarai K, Gauthier W, Hall L, Paudel Krishna P et al (2009) Geographic information systems (GIS) based model of dairy manure transportation and application with environmental quality consideration. Waste Manage 29(5):1634–1643

P-Sphere Project (2017) AgriBigCAT. http://www.psphere-project.com/irta/

Schnase J et al (2017) MERRA analytic services: meeting the big data challenges of climate science through cloud-enabled climate analytics-as-a-service. Comput Environ Urban Syst 61:198–211

Silva S, Alçada-Almeida L, Dias LC (2014) Development of a web-based multi-criteria spatial decision support system for the assessment of environmental sustainability of dairy farms. Comput Electron Agric 108:46–57

Stoate C et al (2009) Ecological impacts of early 21st century agricultural change in Europe—a review. J Environ Manage 91(1):22–46

Teira-Esmatges M, Flotats X (2003) A method for livestock waste management planning in NE Spain. Waste Manage 23(10):917–932

Land-Use Change and CO_2 Emissions Associated with Oil Palm Expansion in Indonesia by 2020

Liselotte Schebek, Jan T. Mizgajski, Rüdiger Schaldach and Florian Wimmer

Abstract The expected increase in palm oil production for food and biofuels has raised large concerns about land-use change and greenhouse gas emissions. The pressure to convert land into oil palm plantations can be widely observed in Indonesia. So far, Indonesia has not been effective in protecting its land resources from this pressure largely because of the weak enforcement of its own policies. Thus understanding the opportunities to improve the policy enforcement in relation to the land resources is critical to design successful strategies for land management in Indonesia. This study simulated land-use changes in Indonesia under three policy scenarios and different projections of palm oil production by 2020. This enabled us to illustrate the effects of the improvements of the policy enforcement on land-use change and CO_2 emission triggered by the growing demand for palm oil. We projected a large increase in deforestation, ranging from 3.06 to 4.89 million hectare if no improvements are made in the policy enforcement, which would result in 194.83–499.89 Mt of CO_2 emission. Better policy enforcement can bring significant mitigation effects in terms of land-use change, as it can reduce deforestation by 50–53%. The effects of enhanced policy enforcement on CO_2 emission from land-use change is even more significant. It can reduce CO_2 emission by 84–87%. Therefore, our results highlighted that the current policies have a substantial potential to protect land resources against the growing pressure on land conversion from palm oil plantations in Indonesia. In order to make such existing policies effective, the government must put considerable efforts on the proper and unconditional enforcement of the policies.

L. Schebek (✉) · J.T. Mizgajski
Technische Universität Darmstadt, 64287 Darmstadt, Germany
e-mail: l.schebek@iwar.tu-darmstadt.de

J.T. Mizgajski
e-mail: j.mizgajski@iwar.tu-darmstadt.de

R. Schaldach · F. Wimmer
Center for Environmental Systems Research, Universität Kassel, 34117 Kassel, Germany
e-mail: schaldach@usf.uni-kassel.de

F. Wimmer
e-mail: wimmer@usf.uni-kassel.de

© Springer International Publishing AG 2018
B. Otjacques et al. (eds.), *From Science to Society*, Progress in IS,
DOI 10.1007/978-3-319-65687-8_5

1   Introduction

Globally, carbon dioxide (CO_2) emissions from land-use change and deforestation account for about 13% of total anthropogenic CO_2 emissions (Friedlingstein et al. 2010). Thus, land-use change and deforestation are significant contributors to climate change. Land-use change, in general, denotes every change in the use of an area. Depending on the amount of carbon stored in the soil and vegetation, and the subsequent land use, it causes a transfer of carbon stock into the atmosphere or carbon uptake from the atmosphere. In the last few decades, Indonesia was a frontrunner in deforestation (Margono et al. 2014). One of the most important factors triggering deforestation in this country was the expansion of palm oil plantations associated with growing demand on crude palm oil (CPO) (Busch et al. 2015). Taking into the government's long-term plans for development of oil palm production (Indonesian Ministry for Economic Affairs 2011), finding ways of mitigating the palm oil industry impact on deforestation in Indonesia is of vital importance.

In our research, we used scenario analysis to investigate the effectiveness of potential strategies in the reduction of carbon flows from land-use change triggered by palm oil expansion in Indonesia. Our approach goes beyond the existing studies examining the impact of the future measures for reducing land-use change in Indonesia (Van der Laan et al. 2016; Austin et al. 2015) in two aspects: First, we employed a land-use change model—LandSHIFT—by which land-use changes can be allocated spatially to the entire territory of Indonesia. Second, our scenarios examine the role of agricultural performance and other measures intended in current Indonesian policy for the protection of forest and peat areas.

2   Methods

Modelling of land-use change The LandSHIFT model was used to calculate spatial and temporal land-use change in Indonesia due to the cultivation of food and energy crops, grazing, and urbanization (Lapola et al. 2010; Schaldach et al. 2011). The model is based on the concept of land use systems (Turner et al. 2007) and couples components that represent the respective anthropogenic and environmental sub-systems. As such, it should be regarded as a system method. In our case study, land-use change was simulated on a raster with a cell size of 5 arcminutes, which is about 9×9 km at the Equator. Each cell was assigned to an Indonesian region (regional level). The regional level was used to define drivers of land-use change, which included population growth, crop and livestock production, crop yield improvements due to technological change as well as information on protected areas. Cell-level information included land use type, human population density, and a set of parameters that describe the landscape characteristics (e.g. terrain slope, potential yields, road infrastructure) and land use restrictions (e.g. protected areas, peatland). During the simulation, LandSHIFT translated the regional model drivers (e.g. crop production) into spatial land use patterns. At the beginning of every time step, the suitability of each raster cell for the different land use types was determined based on the cell-level information. Thereafter, the model used region-level data to determine and allocate the land needed for each crop type, pasture, and settlement in the most suitable cells. The model results are raster maps that depict the spatial and temporal patterns of land-use change until 2020.

Input data LandSHIFT was initialized with a gridded historic land use map (hereafter referred to as base map) representing the year 2010, which combines MODIS land-cover data with additional information on the spatial distribution of crop types. The size of the grid cells is 5 arc-minutes. Human population density for each cell of the base map was derived from the Global Rural-Urban Mapping Project (GRUMP) v1 data set (Balk et al. 2006). Moreover, the model input comprised different spatial datasets used for the suitability analysis. Grid level information on terrain slope was based on data from the Agro-Ecological Zones (AEZ) project (Fischer et al. 2000), while the river network density was calculated as the line density of streams per cell based on the HydroSHEDS 15-second vector data (Lehner et al. 2006). Information on road infrastructure was obtained from the Global Roads Open Access Data Set (gROADSv1) (Center for International Earth Science Information Network—CIESIN—Columbia University and Information Technology Outreach Services—ITOS—University of Georgia 2013). Crop yields and rangeland productivity were calculated by LPJmL for the current climate conditions, defined by the reference period of 1971–2000. For this purpose, data was taken from the CRU TS 2.1 gridded dataset for monthly precipitation, air temperature, cloud cover, and frequency of wet days (Mitchell and Jones 2005). The location of protected areas was derived from the World Database on Protected Areas (IUCN and UNEP-WCMC 2014). The data on location of peatlands and the areas under a moratorium on granting of new licenses and improvement of natural primary forest and peatland governance (hereafter referred to as moratorium on deforestation) were provided by Center for International Forestry Research (CIFOR).

Calculation of CO$_2$ emissions from land-use change Land-use change within the simulation period was determined by analyzing scenario-specific raster maps for 2010 and 2020 generated by LandSHIFT. Land-use change occurred if the land use type of a cell in 2020 was different from the land use type of the cell in 2010. The calculation of CO$_2$ emissions from land-use change (e_l) was done using the following formula:

$$e_l = (CS_R - CS_A) * F * \frac{1}{Y} * \frac{1}{P} - e_B$$

where

e_l annualized emissions from carbon stock change due to land-use change
           $[gCO_2 \ MJ^{-1}]$

CS_R carbon stock in the soil and vegetation associated with reference land use
           $[tC \ ha^{-1}]$

CS_A carbon stock in the soil and vegetation associated with actual land use
           $[tC \ ha^{-1}]$

F   factor for the conversion of C to CO_2 (default = 3.664)

Y   annualizing of carbon stock changes over a 20 year period

P   feedstock productivity $[MJ \ ha^{-1} \ a^{-1}]$

e_B Bonus for re-cultivation of degraded areas (29 $gCO_2 \ MJ^{-1}$).

In our study, CS_R is the carbon stock in 2010, while CS_A represents the carbon stock in 2020. The calculation took into account the carbon stocks in the soil and vegetation. Since the area for which a carbon stock was calculated should have similar conditions in terms of land use, soil and climate, CS_R and CS_A were determined for each cell of the corresponding raster map separately. Cell-level information regarding soil and climate was derived from the *JRC Support to Renewable Energy Directive* project (Joint Research Centre 2011). The *guidelines for the calculation of land carbon stocks* (European Commission 2010) provided land use, soil and climate type dependent default values for soil organic carbon and vegetation carbon stocks (above and below ground). The calculation of carbon stocks for mineral soils and organic soils was consistent with the IPCC Tier 1 methodology.[1] After the carbon stocks for 2010 and 2020 were determined, the annualized GHG emissions from land-use change (e_l) were computed for each cell. To obtain the change in carbon stocks during the simulation period, CS_A was subtracted from CS_R. Then, the yearly emissions related to these carbon stock changes were calculated for a time frame of 20 years by allocating it in 20 equal parts to each year. This was due to the fact that some emissions occur during the conversion process itself and others over a long period of time after the conversion.

Scenario analysis The scenarios used in analysis describe progress in policies influencing land-use change in Indonesia between 2014 and 2020. Accordingly, each scenarios represent different strategy to land governance: Business as Usual (BaU), Medium Governance (MGov), and Best Governance (BGov). Information for scenario development was derived from an investigation on the status and prospects of land-use change in Indonesia based on a comprehensive evaluation of relevant literature and on expert interviews carried out in Indonesia. The scenarios were implemented in the LandSHIFT modelling framework through five key factors influencing the relationship between palm oil production and land-use change in Indonesia (Table 1).

---

[1]See IPCC (2006) *Guidelines for National Greenhouse Gas Inventories, Volume 4: Agriculture, Forestry and Other Land Use* for more information on IPCC Tier 1, 2 and 3 methodologies.

Table 1 Characteristics and transposition method of key factors in scenario analysis

Key factor	Transposition method into LandSHIFT	Scenarios		
		BaU	MGov	BGov
Moratorium on deforestation	Constraints on land use-change within the moratorium borders based on the spatial data provided by CIFOR	Inadequate enforcement	Improved enforcement	Full enforcement
Protection of peat deeper than 3 m	Constraints on land use-change of peatlands deeper than 3 m based on to the spatial data provided by CIFOR	No protection; an increase of suitability of peatland for oil palm production	Limited protection	Full protection
Preservation of protected areas	Constraints on land use-change within protected areas based on the spatial data provided by UNEP-WCMC (IUCN and UNEP-WCMC 2014)	Inadequate enforcement	Improved enforcement	Perfect enforcement
Security of land tenure rights	Constraints on land use-change within land inhabited by indigenous people based on to the spatial data provided by CIFOR	Highly insecure land tenure rights	Moderately insecure land tenure rights	Secure land tenure rights
Agricultural productivity in palm oil sector	Change in oil palm yield in fresh fruits bunches per hectare	Low productivity increase	Medium productivity increase	High productivity increase

Quantities of palm oil production by 2020 were set independently of the scenarios in order to study their effects in the presence of different CPO production levels. Projections of quantities of palm oil produced in Indonesia were specified for three cases: "HIGH" ($\sim$40 Mt), "MEDIUM" ($\sim$35 Mt) and "LOW" ($\sim$32 Mt) production.

The scenarios and quantities of palm oil produced were combined with each other pairwise, leading to 9 combinations that formed the basis for the LandSHIFT simulation runs. All combinations were analyzed against the background of the assumptions in the current developments in agricultural production, including food crops (excluding oil palm) and livestock. These assumptions were the same for all scenario combinations and were based on linear extrapolation of the trend for the period 2008–2013 reported by the Food and Agriculture Organization (FAO) (OECD/FAO 2014).

3 Results

The outcomes of the simulation experiments are presented as conversions between the following land use types: forest, forest on peatland, oil palm, and other food crops. Figure 1 depicts the resulting area of land-use change between 2010 and 2020 as well as the respective CO_2 emissions for the different scenarios, ordered according to HIGH, MEDIUM, and LOW CPO production levels. Figure 2 presents the calculated maps for the scenarios and the base map. The BaU scenario was characterized by the highest amount of land-use change. In combination with a HIGH palm oil production level, 2.29 million ha of forest was converted to oil palm plantations and 2.60 million ha to other types of cropland. Almost 40% of the total forest area converted was located on peat soils. A reduction in the palm oil production to the MEDIUM and LOW levels led to a decrease in forest clearing for new oil palm plantation to 1.23 million ha and 0.46 million ha respectively, while the proportion of peat and non-peat areas remained almost unchanged. At the same time, forest conversion to cropland remained at the same order of magnitude as in the HIGH production scenario because the amount of food crop production is similar in all scenarios. The resulting total CO_2 emission for the BaU scenario amounted to 500 Mt for the HIGH production level. The largest share came from the establishment of new palm oil plantations, which was responsible for 72.2% of the total emission. As a result of LOW CPO production, annual CO_2 emissions substantially decreased to 314 Mt and 195 Mt under the MEDIUM and LOW production levels respectively. Under the MEDIUM production level, forest

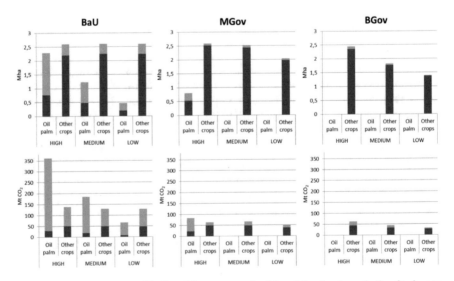

Fig. 1 Conversion of forest (*black*) and forest on peatland (*gray*) to oil palm and other food crops [million ha] until 2020 and related annual CO_2 emissions [Mt] in Indonesia under the different scenarios (BaU/MGov/BGov) and palm oil production levels (HIGH/MEDIUM/LOW)

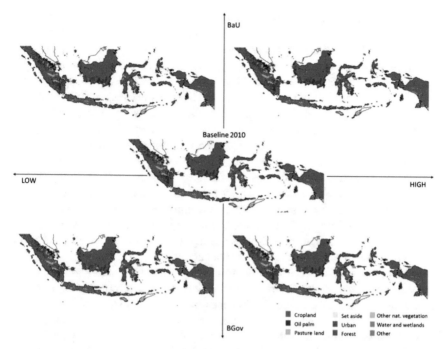

Fig. 2 Simulated land-use change until 2020 in Indonesia for the extreme combination of scenarios (BaU/BGov) and palm oil production levels (HIGH/LOW)

conversion to palm oil plantations was still the main source of emissions, while under the LOW production level, forest conversion to cropland became the dominant emission trigger. Still, most of the emissions stemmed from deforestation located on peatlands (78% for MEDIUM, 71% for LOW).

Forest conversion in the MGov scenario under the HIGH palm oil production level amounted to 3.39 million ha; however, only 0.80 million ha were attributed to new oil palm plantations, while the majority (2.59 million ha) was caused by the expansion of cropland. The minor role played by oil palm in deforestation can be explained by the oil palm yield increase, which compensated for the production increase. Due to the more efficient protection policy for peatland, only 10% of land-use change was located on peat soils. In consequence, the total CO_2 emission from land-use change amounted to 148 Mt per year the HIGH production level. Still, 51% of this (total) emission occurred due to the conversion of forest on peatlands. For the MEDIUM and LOW production levels in the MGov scenario, there was no conversion of forest to oil palm plantation as the oil palm yield increase fully compensated for the production increases. At the same time, cropland expansion decreased to 2.5 million ha and 2.03 million ha under the MEDIUM and LOW production levels respectively. The yield levels of other crops were not changed in the MGov scenario. This development can be explained by the lower competition with palm oil plantations for new land, allowing farmers to access more fertile soils where higher yields can be achieved. As a result of the reduced deforestation, the associated CO_2 emissions decreased to 67.34 Mt and 51.13 Mt under the MEDIUM and LOW production levels respectively, which can generally be attributed to cropland expansion. Under both production levels, the majority of the newly cultivated croplands were located in Sumatra (MEDIUM: 1.67 million ha; LOW: 1.37 million ha). These land-use changes were responsible for 67 and 65% of the total CO_2 emission in the MEDIUM and LOW scenarios. With regards to the regional focus, under the HIGH production level of the MGov scenario, deforestation hot spots were similarly located in Sumatra and Kalimantan.

In the BGov scenario, the projected yield increases were high enough that all production scenarios could be realized on existing plantations, meaning that no additional forest land was converted to palm oil plantations. Consequently, the simulated land-use changes and related CO_2 emissions were only driven by the expansion of cropland required for food production. In comparison with the other scenarios, this area expansion was significantly lower as there was no competition for high productive land with palm oil plantations.

A comparison of the results of land-use change and CO_2 emissions among the different scenarios at the same production level makes it possible to track the influence of yield increase versus policy measures for land protection. In the case of HIGH CPO production level, land-use change was reduced in MGov by 30% and in BGov by 49% compared to the BaU scenario. These numbers display exclusively the influence of yield increase. For the same CPO production level, reductions in CO_2 emissions compared to BaU amounted to 70% in MGov and 87% in BGov, reflecting the combined influence of yield and policy measures.

Figure 2 shows the simulated land-use maps. In the BaU scenario, the regions of Sumatra and Kalimantan were hot-spots of deforestation, both accounting for almost 87% of total forest conversion and 96% of total CO_2 emission. Furthermore, in all scenarios, Sumatra was the region most affected by land-use change, with more than a half of all conversions. The large availability of land suitable for agriculture and well-developed infrastructure fostered the agriculture expansion on the island. A remarkable conversion also took place in Java, Kalimantan, and Sulawesi. Papua and other remaining regions were least affected by land-use change, due to poorly-developed infrastructure and relatively small areas suitable for agriculture. The biggest potential impact of the considered policy and agricultural measures was in Kalimantan. In this region, the change of BaU to BGov scenario reduced land-use change by 70% for HIGH CPO production level. The same scenario change in Sumatra and Papua caused 45% and 40% reductions of land-use change respectively. The high impact of scenario shifts on preventing land conversion in these regions was linked to the fact that large areas of land were already protected, but the protection measures were weakly enforced.

Geographically, Sumatra was the dominant source of CO_2 emissions in all scenarios due to large areas converted. Substantial emissions occurred also in Kalimantan in the BaU and MGov scenarios. Nevertheless, improvement of policy measures according to the MGov and BGov scenarios caused important reductions in CO_2 emission compared to the BaU scenario in these regions.

4 Discussion

Policy implications As pointed out by Carlson et al. (2012), in the period from 1990 to 2010, Indonesia experienced an expansion of palm oil plantation by 600% to 7.8 million ha. Almost 90% of this expansion occurred in Sumatra and Kalimantan. Our results confirm that the conversion of forest to oil palm plantation and other arable land will continue to account for a significant share of land-use change in Indonesia if no substantial change in policy enforcement takes place. The growing oil palm production will generate particularly strong pressure on land as long as there are no major improvements in CPO yield and in the effectiveness of land protection. The improvements of this kind, depicted in MGov and BGov scenarios, can effectively mitigate the pressure of the growing palm oil production. Firstly, land-use change can be reduced by increasing CPO yield per ha (technological change). The increase in yield is able to compensate for the additional land demanded by the growing oil palm production. Results from modelling showed that higher productivity can reduce the area of oil palm plantation, leaving free space for arable land expansion, thus lowering the pressure on forest. Secondly, land protection measures concentrated on peatland, leading to a reduction in the conversion of peat areas. As a consequence, land-use change to oil palm production was reduced to zero, and the conversion of peat areas decreased significantly as a result of better policy performance as depicted in MGov and BGov scenarios.

Land-use change of forest to other arable land still took place; however, this affected carbon flows much less. Generally, in comparison with land conversion, CO_2 emissions from land-use change in Indonesia were much more sensitive to the changes in scenarios. The range of the results across scenarios spreads from 33 to 502 Mt of CO_2 for the whole of Indonesia.

As expected, the highest CO_2 emission was caused by the conversion of peatland, which took place extensively in the BaU scenario. It can be stated that weak peatland protection caused high emission increase as a direct response to the increase in palm oil production. CO_2 emission in the MGov and BGov scenarios were respectively 3–4 and 7 times lower than in the BaU scenario when the same levels of palm oil production were considered. This was because of the shrink in oil palm plantations as a consequence of CPO yield improvement, which in conjunction with the protection of peatland and other improvement measures had a significant impact on CO_2 emission savings. Consequently, as a policy implication, these findings show that, while at present palm oil expansion is directly linked to a rise in CO_2 emissions, better policy enforcement may enable significant reductions of emissions even while increasing CPO production.

Comparison with other studies Our study is the first study that attempts to investigate the impact of policy enforcement on land-use change pressure from palm oil, using land-use change model. In contrast to other studies dealing with this topic (e.g. Afriyanti et al. 2016; Van der Laan et al. 2016), our results provide information on the spatial impact of land-use change mitigation measures. Moreover, the spatial distribution of the results is determined by validated parameters of the model, enabling the simulation of probable land-use change patterns. Another advantage of conducting scenario analysis by means of land-use change models, like LandSHIFT, is that it becomes possible to take into account the competition for land among palm oil, other expanding crops, and infrastructure. The main limitation of our study is that it did not take into account the potential of degraded areas to provide low-carbon and underutilized land for palm oil expansion. We believe that activating this option could lower the pressure on forest and peatland by the palm oil industry. However, currently, there is insufficient data to examine the potential of degraded land in all territories of Indonesia. The inclusion of this feature in land-use change models dealing with palm oil expansion in Indonesia is highly recommended.

References

Afriyanti D, Kroeze C, Saad A (2016) Indonesia palm oil production without deforestation and peat conversion by 2050. Sci Total Environ 557–558:562–570. doi:10.1016/j.scitotenv.2016. 03.032

Austin KG, Kasibhatla PS, Urban DL, Stolle F, Vincent J (2015) Reconciling oil palm expansion and climate change mitigation in Kalimantan, Indonesia. PLoS ONE 10(5):e0127963. doi:10. 1371/journal.pone.0127963

Balk DL, Deichmann U, Yetman G, Pozzi F, Hay SI, Nelson A (2006) Determining global population distribution: methods, applications and data. Adv Parasitol 62:119–156

Busch J, Ferretti-Gallon K, Engelmann J, Wright M, Austin KG, Stolle F, Turubanova S, Potapov PV, Margono B, Hansen MC, Baccini A (2015) Reductions in emissions from deforestation from Indonesia's moratorium on new oil palm, timber, and logging concessions. Proc Natl Acad Sci 112(5):1328–1333. doi:10.1073/pnas.1412514112

Carlson KM, Curran LM, Ratnasari D, Pittman AM, Soares-Filho BS, Asner GP, Trigg SN, Gaveau DA, Lawrence D, Rodrigues HO (2012) Committed carbon emissions, deforestation, and community land conversion from oil palm plantation expansion in West Kalimantan, Indonesia. Proc Natl Acad Sci 109(19):7559–7564. doi:10.1073/pnas.1200452109

Center for International Earth Science Information Network—CIESIN—Columbia University, Information Technology Outreach Services—ITOS—University of Georgia (2013) Data from Global Roads Open Access Data Set, Version 1 (gROADSv1). NASA Socioeconomic Data and Applications Center (SEDAC). doi:10.7927/H4VD6WCT

Commission European (2010) Commission decision of 10 June 2010 on guidelines for the calculation of land carbon stocks for the purpose of Annex V to Directive 2009/28/EC. OJL 2010(151):19–41

Fischer G, van Velthuizen HT, Nachtergaele FO (2000) Global agro-ecological zones assessment: methodology and results. IIASA, Laxenburg

Friedlingstein P, Houghton RA, Marland G, Hackler J, Boden TA, Conway TJ, Canadell JG, Raupach MR, Ciais P, Le Quere C (2010) Update on CO_2 emissions. Nat Geosci 3(12):811–812

Indonesian Ministry for Economic Affairs (2011) Masterplan for acceleration and expansion of Indonesia economic development 2011–2025

IUCN, UNEP-WCMC (2014) Data from the world database on protected areas (WDPA). https://www.protectedplanet.net/

Joint Research Centre (2011) European Soil Data Centre (ESDAC). http://eusoils.jrc.ec.europa.eu/projects/renewable-energy-directive

Lapola DM, Schaldach R, Alcamo J, Bondeau A, Koch J, Koelking C, Priess JA (2010) Indirect land-use changes can overcome carbon savings from biofuels in Brazil. Proc Natl Acad Sci 107 (8):3388–3393. doi:10.1073/pnas.0907318107

Lehner B, Verdin K, Jarvis A (2006) HydroSHEDS technical documentation, version1.0. http://hydrosheds.cr.usgs.gov/webappcontent/HydroSHEDS_TechDoc_v10.pdf

Margono BA, Potapov PV, Turubanova S, Stolle F, Hansen MC (2014) Primary forest cover loss in Indonesia over 2000–2012. Nat Clim Change 4(8):730–735. doi:10.1038/nclimate2277

Mitchell TD, Jones PD (2005) An improved method of constructing a database of monthly climate observations and associated high-resolution grids. Int J Climatol 25(6):693–712

OECD/FAO (2014) OECD-FAO agricultural outlook 2014. OECD Publishing, Paris

Schaldach R, Alcamo J, Koch J, Kölking C, Lapola DM, Schüngel J, Priess JA (2011) An integrated approach to modelling land-use change on continental and global scales. Environ Model Softw 26(8):1041–1051. doi:10.1016/j.envsoft.2011.02.013

Turner BL, Lambin EF, Reenberg A (2007) The emergence of land change science for global environmental change and sustainability. Proc Natl Acad Sci 104(52):20666–20671. doi:10.1073/pnas.0704119104

Van der Laan C, Wicke B, Verweij PA, Faaij APC (2016) Mitigation of unwanted direct and indirect land-use change—an integrated approach illustrated for palm oil, pulpwood, rubber and rice production in North and East Kalimantan, Indonesia. GCB Bioenergy. doi:10.1111/gcbb.12353

Application of the Forgotten Effects Theory for Assessing the Public Policy on Air Pollution of the Commune of Valdivia, Chile

Erna Megawati Manna, Julio Rojas-Mora and Cristian Mondaca-Marino

Abstract Environmental public policy is challenged by its complexity and uncertainty as manifested in Wicked Problems. Incentives were considered to encourage behavioural change in the public policy on air pollution in Valdivia, Chile. However, air pollution is still increasing due to extensive usage of firewood. This research aims to assess the policy within the identified incentives and behaviours by applying the Forgotten Effects theory. Empirical evidence of Forgotten Effects in the policy is provided by using data from key informant interviews. A major finding of the research is that a subsidy on transportation has no effect on reducing air pollution in residential areas. It also suggests that if people improve thermal insulation of their houses in the commune of Valdivia may contribute to a major reduction of air pollution rather than improving the quality of firewood. Education, Research and Development may also play an important role to encourage behavioural change. The latter will also contribute to the improvement of overall energy efficiency and thereby lower emissions.

Keywords Air pollution · Behaviour · Complexity · Environmental planning · Firewood · Forgotten effects · Incentive · Planning · Public policy · Uncertainty · Wicked problems

E.M. Manna (✉)
Bidang Perencanaan, Pengendalian, dan Evaluasi pada Badan Perencanaan,
Penelitian, dan Pengembangan Daerah Kabupaten Sumba Barat Daya
(Planning, Controlling, and Evaluation Division at Regional Planning, Research,
and Development Board of Southwest Sumba), Southwest Sumba, Indonesia
e-mail: mega_ern@yahoo.com

J. Rojas-Mora
Escuela de Ingeniería Informática en Universidad Católica de Temuco (The School
of Informatic Engineering at Catholic University of Temuco), Temuco, Chile
e-mail: jrojas@inf.uct.cl

C. Mondaca-Marino
Facultad de Ciencias Económicas y Administrativas en Universidad Austral de Chile
(The Faculty of Economics and Management Science at Austral University of Chile),
Valdivia, Chile
e-mail: cristianmondaca@uach.cl

© Springer International Publishing AG 2018
B. Otjacques et al. (eds.), *From Science to Society*, Progress in IS,
DOI 10.1007/978-3-319-65687-8_6

1   Introduction

Air pollution has become a major environmental problem in Chile. Especially during the winter season when firewood is used for heating (Schueftan and González 2013). The commune of Valdivia in the central-southern area of Chile in no exception. The local government of Valdivia has drafted a decontamination plan (*PDA, Plan de Descontaminación Atmosférica para la comuna de Valdivia, en la Resolución Exenta N° 839, de 24 de Agosto de 2015*) which focuses on reducing emissions from firewood burned by the residential sector targeting four strategic areas: (i) people improve thermal insulation of houses, (ii) people improve firewood combustion equipment efficiency and its components, (iii) people improve firewood quality and availability of other combustibles, and (iv) education and awareness in the community.

This plan considers three key incentive programmes to control contamination, i.e. (i) a subsidy for stove replacement which mainly focuses on heater and cooking appliances, (ii) the thermal insulation and refurbishment of housing, and (iii) the implementation of a private-state certification system for retail firewood. However, according to Schueftan and González (2015) policies so far have not had the desired effect on lowering air pollution.

The failure of policies to reduce hazardous emissions, in line with the concept of "Wicked Problems" include nearly all public policy issues on environmental and planning problems. The concept of Wicked Problems was introduced by Rittel and Webber (1973) as a new approach contrasting common approaches at the time which tackled problems in planning and design of large-scale systems in a corrective rationalist approach. Wicked Problems are ill-defined, ambiguous and associated with strong moral, political and professional issues. Since they are strongly stakeholder-dependent, there is often little consensus on problem definition. Above all, Wicked Problems are evolving constantly (Ritchey 2013). In addition, Wicked Problems are characterized by a high degree of uncertainty and a profound lack of agreement on common values (Allen and Gould 1973). Rittel and Webber (1973) distinguish ten properties of planning-type problems that planners need to be aware of and need to take into account during the planning process.

In the case of the environmental public policy in Valdivia, one of the Wicked Problem properties that could fit for this issue is the fact that an immediate and ultimate test of a solution does not exist. After the implementation period, waves of consequences over an extended period of time may be generated (Rittel and Webber 1973).

The "Forgotten Effects" theory (Kaufmann and Gil Aluja 1988) allows for the modelling of Wicked Problems providing a very useful tool for the evaluation of systemic complexities that arise in public policies in general, and those that affect the environment in particular.

The objectives of this paper are twofold. Firstly, we introduce the use of the Forgotten Effects theory as a new methodology for the identification of indirect incentives in public policy. Secondly, we apply this methodology to the ex-ante evaluation of the PDA of Valdivia.

2   Methodological Approach

A questionnaire was prepared which functioned as a main data gathering instrument on the basis of the analysed policy document. The questionnaire consisted of open questions to explore the opinions of key informants related to the policy. Furthermore, three matrices were filled by the key informants based on criteria provided in the questionnaire. These matrices were covered in-depth hereafter. The selection of key informants for data collection was based on the so-called "snowball" sampling method because it focuses on practitioners (Bernard 2006). The number of key informants for the topic was relatively small, which is why it was a straightforward task to identify key informants. The data were then transcribed to Microsoft Excel for further analysis. For the final analysis, the R scripting software was used as the main statistical tool for testing and applying the Forgotten Effects theory, with a minor analysis based on graph centrality.

To design a public environmental policy that faces complexity, economic incentives are usually introduced. Incentive is all actions implemented by public policy to reward or penalize behavioural change of population (Chomitz and Birdsall 1990). From this definition, it is clear that there is a direct relationship between incentives and behaviours. Incentives are not only comprised of subsidies or money given to encourage behavioural change but also include prohibitions, standards or regulations to assist the implementation of a policy.

Table 1 provides the list of incentives that were identified on the basis of PDA that are meant to influence behaviours or to reach the desired outcomes. Each identified incentive is the result of the policy review, discussions, and key informant interviews and is related to the article numbers in this table. A total of ten key informants from six different institutions which vary from academia and government were interviewed for the policy review. This procedure is deterministic in nature thus the representativeness is only addressed in the organizations to which key informants belong to. They were three persons from Austral University of Chile, one person from *Certificación e Investigación de la Vivienda Austral* (CIVA),[1] one person from *Corporación Nacional Forestal* (CONAF),[2] two persons from *Instituto Forestal* (INFOR),[3] one person from Ministry of Energy, one person from Ministry of Environment, and one person from Municipality of Valdivia. Table 2 shows the list of behaviours based on the four strategic areas that exist in PDA, which are concordant with the strategy of the Ministry of Environment to reduce emissions from the residential sector.

---

[1]Austral Certification and Research of Housing.
[2]National Forest Corporation.
[3]Forestry Institute.

2.1 Forgotten Effects Theory

Economists emphasize that "incentives matter" (Gneezy et al. 2011). The basic "law of behaviour" states that higher incentives will lead to more effort and higher performance (Gneezy et al. 2011). The methodological foundation of Forgotten Effects is comprised of two sets of elements (causes and effects) which can be interpreted as incentives and behaviours. In public policy, each incentive is targeted at modifying a subset of behaviours, in a directed manner. Nevertheless, there are relations between incentives as well as relations between behaviours that are not clear in the final draft of most public policy documents. They, as cables of an old switching board, enable indirect relationships between incentives and behaviours, what Kaufmann and Gil-Aluja called "Forgotten Effects". The concept of incidence is associated with the number of times something happens or develops. The incidence is a highly subjective concept and is usually difficult to measure. However, a thorough analysis may improve reasoned action and decision-making (Brown et al. 2007).

Given a set $A = \{a_i \mid = 1, 2, 3, ..., n\}$ of incentives and a set $B = \{b_j \mid = 1, 2, 3, ..., m\}$ of behaviours an example incidence graph is depicted in Fig. 1. Table 3a shows the incidence matrix based on Fig. 1. The incidence values $\mu(a_i, b_j)$ in this example are either 0 or 1. However, we followed an approach which assigns incidence values $\mu(a_i, b_j)$, Degrees of Truth (Kaufmann and Gil Aluja 1988), of the proposition "a_i has incidence on b_j" from the interval [0, 1] as shown in Table 3b $\{\forall \, (a_i, b_j) \in A \times B, \, \mu \, (a_i, b_j) \in [0, 1]\}$.

Table 1 Identified incentives

Code	Incentives	Article number based on the PDA														
I1	Prohibition	4	5	6	8	9	10	49	53							
I2	Standard (regulation)	7	22	30	31	32	37	39	40	41	42	43	47	51	56	57
I3	Subsidy for heating devices for public services	11														
I4	Subsidy for heating devices for households	12														
I5	Record (inventory, information)	13	15	20	21	25	26	38	46							
I6	Dry firewood	17														
I7	Inspection and audit	18	33	35	55	75	76	77	78	79	80	81	82	83	84	
I8	Improvement and production	19														

(continued)

Table 1 (continued)

Code	Incentives	Article number based on the PDA																
I9	Subsidy for firewood	23	24															
I10	Subsidy for thermal insulation of houses	27	28	29														
I11	Education	34	50	70	71	72	73	74										
I12	Subsidy for boilers	45																
I13	R&D, research and development	35	36	44	48													
I14	Subsidy for transportation	54																
I15	Environmental impact assessment system	58	59	60	61													
I16	Mitigation	62	63	64	66	67	68	69										

Table 2 Identified behaviours

Code	Behaviours
B1	People improve thermal insulation of houses
B2	People improve firewood combustion equipment efficiency and its components
B3	People improve firewood quality and availability of other combustibles
B4	Education and awareness in the community

Fig. 1 Graph incidence of A over B (adapted from Kaufmann and Gil Aluja 1988)

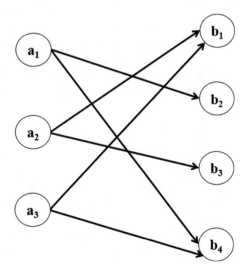

Table 3 (a) Incidence Matrix of A over B, (b) degrees of truth (adapted from Kaufmann and Gil Aluja 1988)

(a)

	b_1	b_2	b_3	b_4
a_1	0	1	0	1
a_2	1	0	1	0
a_3	1	0	0	1

(b)

Degrees of Truth	Label
0	False
0.1	Basically false
0.2	Almost false
0.3	Rather false
0.4	More false than true
0.5	As true as false
0.6	More true than false
0.7	Relatively true
0.8	Almost true
0.9	Practically true
1.0	True

In order to identify second or higher order effects, incidence graphs are generated as $AB = \left(ac_{ij}\right)_{n \times m}$ and $BC = \left(bc_{jk}\right)_{m \times s}$ with a max-min composition of $AC = AB \circ BC$ where $ac_{ik} = \max_{j=1,\ldots,m} \min\left(ab_{ij}, bc_{jk}\right)$, $\forall_i = 1,\ldots,n, k = 1,\ldots,s$. Using the max-min composition, the cumulative first and second order effects of incentives on behaviours are identified by $AB^{1+2} = AA \circ AB \circ BB$. The max-min composition follows the associative rule $(AA \circ AB) \circ BB = AA \circ (AB \circ BB)$. The matrix AB' is calculated using a max-min composition of rows AB and columns BB. Then AB^{1+2} is derived as $AB^{1+2} = AA \circ AB'$. To determine the second order effects AB^2, the first order AB matrix is subtracted: $AB^2 = AB^{1+2} - AB$.

The AB (incentives-behaviours), AA (incentives-incentives), and BB (behaviours-behaviours) matrices were derived by information obtained from the key informants. Weighting of the edges incentives-behaviours were obtained from the evaluation by key informants as they provided an interval that used as an input for sensitivity analysis. A weight was randomly selected from the interval provided by each key informant for each iteration of this analysis. For instance, incidence value of incentive *I1* to behaviour *B1* is selected from a key informant that has range between 0.5 and 0.8 (see Table 3b). Then, the mean value was used for the calculation of the Forgotten Effects. After the data analysis, key informants were then asked to validate the results.

2.2 Graph Centrality

Besides the Forgotten Effects theory, which intends to identify indirect or second order effects, centrality concepts from the graph theory were applied in order to evaluate direct or first order effects. Graphs consist of nodes and edges which connect nodes. A graph is called "connected" when every node is reachable from any other node. A distance may be associated with each path in a graph; this might be looked at as the number of edges in that particular path. The shortest paths linking a given pair of nodes are called "geodesics" (Freeman 1979).

The relative importance of a node or edge within a graph is determined using centrality measures. Closeness and PageRank were applied for centrality measures. Closeness is defined as the overall mean of the shortest path between a node and all other nodes reachable from it (Noori 2011). The more central a node is, the lower its total distance from all other nodes. PageRank is an algorithm used by Google Search to rank search engine results. It was named after Lawrence Page, the Google founder. PageRank can be thought of as a model of user behaviour. PageRank works by counting the number and quality of links to a node to determine a rough estimate of how important it is. The underlying assumption is that more important nodes are likely to receive more links from other nodes (Brin and Page 1998).

Methodological foundations of closeness and PageRank centrality measures are defined hereafter. Three schemas were carried out using minimum (Schema 1), average (Schema 2), and maximum (Schema 3) values based on expert opinions. For each schema, the values were calculated and an adjacency matrix G was constructed that summarizes the opinion of key informants in the following way:

$$G = \left[g_{i,j}\right]_{(n+m)\times(n+m)} = \begin{bmatrix} 0 & aa_{1,2} & \cdots & aa_{1,n} & ab_{1,1} & ab_{1,2} & \cdots & ab_{1,m} \\ aa_{2,1} & 0 & \cdots & aa_{2,n} & ab_{2,1} & ab_{2,2} & \cdots & ab_{2,m} \\ \vdots & \vdots & \ddots & \vdots & \vdots & \vdots & \ddots & \vdots \\ aa_{n,1} & aa_{n,2} & \cdots & 0 & ab_{n,1} & ab_{n,2} & \cdots & ab_{n,m} \\ 0 & 0 & \cdots & 0 & 0 & bb_{1,2} & \cdots & bb_{1,m} \\ 0 & 0 & \cdots & 0 & bb_{2,1} & 0 & \cdots & bb_{2,m} \\ \vdots & \vdots & \ddots & \vdots & \vdots & \vdots & \ddots & \vdots \\ \vdots & 0 & \cdots & 0 & bb_{3,1} & bb_{3,2} & \cdots & 0 \end{bmatrix}$$

The adjacency matrix G has the following characteristics: the top-left side consists of matrix AA (incentives-incentives), the top-right side represents matrix AB (incentives-behaviours), the bottom-left side consists of zeros, and the bottom-right side represents matrix BB (behaviours-behaviours). The diagonal of the matrix is composed by zeros, i.e. the relationship of a node with itself in the graph.

For each graph, the closeness and PageRank centrality measures were calculated. PageRank centrality is especially appropriate for evaluating behaviours, as the algorithm states that the most important nodes are the ones that receive many links from important nodes. Closeness shows that either incentives or behaviours should be considered important. Nevertheless, in the context of this work, more importance is paid to incentives than to behaviours. For closeness centrality, a distance matrix D is calculated as the complement of the original matrix which is D = 1 − G.

The reason behind this is that in the adjacency matrix, the values are directly related to incidences (the bigger the value the stronger the incidence). As it is of interest to see how close a node is to the rest, the distance as the complement of the incidence needs to be determined, i.e. the smaller the distance (the higher the incidence) the better.

3   Results

3.1   Direct Effects (First Order Effects)

Three schemas were developed to calculate closeness and PageRank to determine first order effects. Minimum (Schema 1), average (Schema 2) and maximum (Schema 3) values based on expert opinions were determined and are presented in Table 4 for closeness and PageRank for all identified incentives and behaviours.

Table 4 shows that incentives in most cases have bigger values than behaviours in closeness, while for PageRank the values are of similar magnitude for both incentives and behaviours. Since PageRank determines the number and quality of nodes in order to estimate the importance of a node, the most important nodes are likely to receive more links from least important nodes. Note that incentives are origins (causes) and behaviours are destinations (effects).

Behaviours in closeness show more or less the same value (about 0.05), while in PageRank, the first behaviour *B1 (People improve thermal insulation of houses)* shows bigger values than the other behaviours. The opinion of key informants was that all incentives and behaviours will point to this behaviour.

Table 4 Minimum, average, and maximum values of closeness and PageRank

		Minimum (schema 1)		Average (schema 2)		Maximum (schema 3)	
		Closeness	PageRank	Closeness	PageRank	Closeness	PageRank
Incentives	I1	0.097	0.053	0.112	0.054	0.132	0.055
	I2	0.122	0.049	0.141	0.049	0.168	0.049
	I3	0.086	0.057	0.097	0.058	0.115	0.059
	I4	0.096	0.054	0.109	0.054	0.126	0.055
	I5	0.104	0.055	0.119	0.055	0.141	0.056
	I6	0.111	0.046	0.127	0.046	0.148	0.046
	I7	0.108	0.049	0.124	0.049	0.145	0.048
	I8	0.109	0.047	0.127	0.047	0.153	0.047
	I9	0.091	0.055	0.103	0.056	0.119	0.057
	I10	0.097	0.054	0.110	0.054	0.128	0.054
	I11	0.131	0.048	0.152	0.048	0.173	0.048
	I12	0.090	0.055	0.102	0.055	0.119	0.056
	I13	0.124	0.049	0.145	0.048	0.176	0.048
	I14	0.065	0.066	0.070	0.068	0.077	0.070
	I15	0.088	0.054	0.098	0.054	0.109	0.055
	I16	0.099	0.046	0.111	0.046	0.127	0.045
Behaviours	B1	0.058	0.045	0.058	0.045	0.059	0.045
	B2	0.058	0.036	0.058	0.034	0.059	0.032
	B3	0.058	0.034	0.058	0.033	0.059	0.030
	B4	0.059	0.039	0.060	0.037	0.060	0.036

An important result from Table 4 can be derived from the closeness section, where incentives *I11* (*Education*) and *I13* (*R&D, Research and Development*) show the highest values compared to the other incentives. These two incentives are complementary to each other. Education is believed to be the most important means to encourage behavioural change. Key informants affirm that if the government invests money in education then this will result in more educated citizens, who in turn will support the program in reducing emissions from the residential sector. Key informants are also confident that promoting and strengthening community processes through dissemination, awareness, training and education by addressing various actions and initiatives will have a great impact on the clean air. Likewise, *R&D* is also considered as a main incentive that the government should prioritize and invest in. In this case, it is stated that government should invest money on studies how to build houses with well-established thermal insulation which is in accordance with energy efficiency. On the other hand, *I14* (*Subsidy for transportation*) shows the lowest values in closeness, which clearly indicate that this incentive is not going to effect on behavioural change for this policy.

3.2 Indirect Effects (Second Order Effects)

Table 5a and b show the summary of minimum, average, and maximum values of matrix AB (incentives-behaviours) and BB (behaviours-behaviours) derived from all key informants, respectively. Matrix AA (incentives-incentives) is by far too large to be contained in this paper but it is available and can be downloaded on Dropbox.[4] From Table 5a, it can clearly be seen that incentive *I14* (*Subsidy for transportation*) has the lowest effect on all behaviours.

No Forgotten Effects could be identified in these data. The reason for this very likely is that the evaluation of direct effects is well above 0.5. Therefore, no indirect affects appeared. Nevertheless, a sensitivity analysis, based on a Monte Carlo simulation with 1000 replicates, was carried out. The sensitivity analysis was carried out by removing two data sets of key informants from two different institutions such Ministry of Environment and INFOR which were estimated to be too optimistic from the other key informants with almost all values ranging from 0.9 to 1.0 ("practically true" to "true" according to the Degrees of Truth in Table 3b).

Table 5c provides the relative frequency of Forgotten Effects resulting from a sensitivity analysis. The result shows that 31%, incentive *I9* (*Subsidy for firewood*) had an indirect effect on behaviour *B1* (*People improve thermal insulation of houses*). The effects of subsidizing firewood should be analysed with care as this might thwart energy effectiveness. However, it seems evident that restricting the use of certain fuels and controlling the use of fuels will result in an improvement of

---

[4]https://www.dropbox.com/s/0i2g35f8389f2a0/Minimum%2C%20Average%2C%20and%20Maximum%20Values%20of%20Matrix%20AA%20%28Incentives-Incentives%29.pdf?dl=0.

Table 5 (a) Minimum, average, and maximum values of matrix AB (incentives-behaviours), (b) minimum, average, and maximum values of matrix BB (behaviours -behaviours), (c) frequency of forgotten effects

(a)

Code	Minimum				Average				Maximum			
	B1	B2	B3	B4	B1	B2	B3	B4	B1	B2	B3	B4
I1	0.34	0.39	0.47	0.43	0.41	0.46	0.54	0.50	0.47	0.52	0.61	0.56
I2	0.69	0.71	0.68	0.67	0.74	0.78	0.73	0.71	0.79	0.84	0.78	0.75
I3	0.27	0.76	0.57	0.43	0.31	0.83	0.64	0.49	0.34	0.89	0.70	0.54
I4	0.29	0.75	0.61	0.45	0.34	0.82	0.68	0.51	0.38	0.88	0.74	0.57
I5	0.34	0.55	0.57	0.67	0.40	0.61	0.65	0.73	0.46	0.66	0.73	0.78
I6	0.39	0.73	0.85	0.59	0.43	0.79	0.90	0.65	0.47	0.85	0.95	0.70
I7	0.72	0.64	0.76	0.56	0.76	0.70	0.82	0.64	0.79	0.76	0.87	0.71
I8	0.50	0.60	0.78	0.57	0.57	0.67	0.84	0.65	0.63	0.74	0.89	0.73
I9	0.18	0.55	0.72	0.46	0.24	0.62	0.77	0.53	0.29	0.69	0.82	0.60
I10	0.88	0.46	0.49	0.67	0.92	0.53	0.56	0.72	0.95	0.59	0.62	0.77
I11	0.77	0.77	0.83	0.95	0.81	0.81	0.88	0.98	0.85	0.85	0.92	1.00
I12	0.32	0.72	0.60	0.36	0.38	0.78	0.67	0.43	0.44	0.83	0.73	0.50
I13	0.83	0.83	0.83	0.86	0.89	0.89	0.89	0.90	0.94	0.94	0.94	0.94
I14	0.10	0.09	0.16	0.31	0.16	0.15	0.21	0.37	0.21	0.20	0.26	0.42
I15	0.47	0.48	0.47	0.41	0.55	0.56	0.53	0.47	0.63	0.63	0.59	0.53
I16	0.43	0.48	0.56	0.57	0.50	0.55	0.63	0.65	0.56	0.61	0.70	0.72

(b)

Code	Minimum				Average				Minimum			
	B1	B2	B3	B4	B1	B2	B3	B4	B1	B2	B3	B4
B1	1.00	0.64	0.54	0.66	1.00	0.71	0.60	0.73	1.00	0.78	0.66	0.79
B2	0.51	1.00	0.73	0.62	0.57	1.00	0.80	0.68	0.63	1.00	0.86	0.73
B3	0.49	0.77	1.00	0.59	0.55	0.84	1.00	0.65	0.61	0.90	1.00	0.70
B4	0.75	0.76	0.77	1.00	0.80	0.81	0.82	1.00	0.85	0.85	0.87	1.00

(c)

From	Through	To	Frequency of appearing
I9	I8	B1	21
I9	I11	B1	7
I9	I10	B1	3
I1	I7	B1	3

thermal insulation of houses. On the other hand, by improving access to certified firewood with subsidies, there will be an improvement on the market of certified firewood. This is which in turn will decrease the need for an improvement of thermal insulation of houses. Instead of dealing with a permanent and long-term solution, government will try to use short-term solution. Here, the proposition of Wicked Problems is clearly evident that there is no immediate and ultimate of solution.

4 Key Findings and Summary

Based on the analysis provided, the research shows that direct effects of the policy *Education* and *R&D, Research & Development* are regarded to have the highest impact (value). This research reveals that incentives to encourage citizen behaviours in *People improve thermal insulation of houses* are very prominent and will thus result on reducing air pollution. *Subsidy for transportation* as an incentive has the lowest impact (value) on reducing air pollution in total as it is not going to have an effect to reduce air pollution in the residential sector. Therefore, it can be stated that *Subsidy for transportation* is a completely inappropriate incentive in this policy considering all other strategies that contemplate to be achieved in Valdivia.

The research shows that Forgotten Effects appear in the policy only when the data of the most optimistic key informants is removed. The most dominant incentive, *Subsidy for firewood*, is identified as the main cause on behavioural change in the context of *People improve thermal insulation of houses* through *Improvement and production*. Again, *People improve thermal insulation of houses* affirmed that it is a very important issue that should be considered in this policy. From the opinions of key informants, we believe that this strategy will affect the reduction of air pollution in Valdivia the most. Therefore, the government should put more effort to create incentives that make people improve thermal insulation of their houses.

Forgotten Effects should be taken into account in policy making for reducing air pollution in Valdivia. In particular, (i) studies on thermal insulation of houses, (ii) encouragement of public participation, (iii) community capacity building, (iv) knowledge management on available technologies, and (v) re-planning of existing strategies should be considered in the future.

The analysis shows that there are Forgotten Effects or important considerations in the public policy that should be prioritized by the government or policy makers. The focus on *Subsidy for firewood* and its certification, and of *Subsidy for transportation* is of lower importance. Even if a subsidy for firewood is provided to citizens, the bad thermal insulation of their houses will only make people to burn more certified firewood, proving it as useless or with no effect at all. In the same sense, while improving the conditions of public transportation will have some effect, the incidence this industry has, at least in Valdivia, is really small compared to other pollution generating problems, like burning wet firewood.

The project also revealed that dealing with complex problems requires a clear process understanding and realistic time frame setting. As a Wicked Problem proposition states, there is no immediate and no ultimate test of a solution to a Wicked Problem. Therefore, a thorough and well-thought through assessment from authorities responsible in decision making is needed. The analysis also indicates that in dealing with complex and uncertain Wicked Problems, mathematical models as presented in this paper can be helpful to deal with these problems. The presented methodology allows to identify Forgotten Effects in policies and provide policy makers with recommendations enhancing them.

References

Allen G, Gould J (1973) Complexity, wickedness and public forests. J Forest 20–23

Bernard R (2006) Research methods in anthropology: qualitative and quantitative approaches

Brin S, Page L (1998) The anatomy of a large-scale hypertextual web search engine (PDF). Comput Netw ISDN Syst 30:107–117. doi:10.1016/S0169-7552(98)00110-X

Brown V, Harris J, Russell J (2007) Tackling wicked problems. doi:10.4324/9781849776530. http://www.apsc.gov.auu

Chomitz KM, Birdsall N (1990) Incentives for small families: concepts and issues. World Bank Res Observer 5:309

Freeman LC (1979) Centrality in social networks conceptual clarification. Soc Netw 1(3):215–239. doi:10.1016/0378-8733(78)90021-7

Gneezy U, Meier S, Rey-Biel P (2011) When and why incentives (don't) work to modify behaviour. J Econ Perspect 25(4):191–210. doi:10.1257/jep.25.4.191

Kaufmann A, Gil Aluja J (1988) Modelos para la investigación de efectos olvidados. [S.l.] Milladoiro

Luce RD, Perry AD (1949) A method of matrix analysis of group structure. Psychometrika 14(2):95–116

Noori A (2011) On the relation between centrality measures and consensus algorithms. In: Proceedings of the 2011 international conference on high performance computing and simulation, HPCS 2011, pp. 225–232. doi:10.1109/HPCSim.2011.5999828

Ritchey T (2013) Wicked problems. Modelling social messes with morphological analysis. Acta Morphologica Generalis 2(1):1–8. doi:10.1016/j.jacr.2013.08.013

Rittel H, Webber MM (1973) Dilemmas Gen Theory Plann 4(2):155–169

Schueftan A, González AD (2013) Reduction of firewood consumption by households in south-central Chile associated with energy efficiency programs. 63:823–832. doi:10.1016/j.enpol.2013.08.097

Schueftan A, González AD (2015) Proposals to enhance thermal efficiency programs and air pollution control in south-central Chile. Energy Policy 79:48–57. doi:10.1016/j.enpol.2015.01.008

Initial Assessment of Air Pollution and Emergency Ambulance Calls in 35 Israeli Cities

Barak Fishbain and Eli Yafe

Abstract Statistical analysis was done for correlations between chronic exposure to ambient air pollution concentrations and emergency ambulance calls in 35 Israeli cities between 2007 and 2012. A second analysis was done between the rate of acute exposure through the rate of extreme events (at least 2 standard errors over the national mean) over the same time period and in the same cities. This study is unique by assessing this association by looking at emergency calls' locations rather than the hospital location. The importance of such study is twofold. First, it facilitates the consideration of environmental conditions in the process of emergency services staffing, allowing for a better utilization of professional personnel and improving overall service. Second, it may shed new light on air pollutants and their correlations to specific medical emergency conditions.

Keywords Air pollution · Environmental health · Medical emergencies · Air pollution exposure

1 Introduction

The health effects of different air pollutants have been well documented (e.g., Beelen et al. 2014; Brauer et al. 2008; Shah et al. 2012; Smith 1987). Carbon Monoxide (CO), originating mostly from anthropogenic activity, typically emissions from transportation (Badr and Probert 1994), was linked with emergency admissions for cerebrovascular diseases (Chan et al. 2006) and myocardial infraction (MI) (Mustafić et al. 2012). Having said that, a comprehensive 6-year

B. Fishbain (✉)
Faculty of Civil and Environmental Engineering,
Technion-Israel Institute of Technology, Haifa, Israel
e-mail: fishbain@technion.ac.il

E. Yafe
Magen David Adom, 60, Yigal - Alon Street, 67062 Tel Aviv, Israel
e-mail: Eliy@mda.org.il

© Springer International Publishing AG 2018
B. Otjacques et al. (eds.), *From Science to Society*, Progress in IS,
DOI 10.1007/978-3-319-65687-8_7

73

study done in Edmonton (Canada) (Villeneuve et al. 2012), showed no statistically significant correlation between ambient levels of CO and hospital visits for stroke, i.e., cerebrovascular accidents. Elevated exposure to CO also showed no effect on premature births (Srám et al. 2005).

Nitrogen oxides (NO, NO_2 and collectively NO_x) arise in the atmosphere largely from the bonding of atmospheric nitrogen and oxygen under high temperature conditions (Badr and Probert 1993). NO_2 ambient levels were found to be correlated to emergency room admissions for hypertension (Guo et al. 2010; Szyszkowicz et al. 2012), asthma aggravation (Koenig 1999) and MI (Mustafić et al. 2012). Weak correlations were found between strokes and NO_2 levels (Villeneuve et al. 2012). A lower prevalence of measured high blood pressure was found in the population exposed to elevated NO_x (Sørensen et al. 2012). NO_2 was associated with increased odds for premature birth only when combined with SO_2 and PM_{10} exposure (Srám et al. 2005).

Tropospheric ozone is a secondary pollutant, originating from complex photo-catalytic interactions between Volatile Organic Compounds (VOCs), NO_x and other atmospheric constituents (Jenkin and Clemitshaw 2000). While exposure to higher levels of ozone was found to be a risk factor for hypertension (Dong et al. 2013) and cerebrovascular diseases (Chan et al. 2006), other studies concluded it did not play a role in increasing risk for MI (Mustafić et al. 2012) and stroke (Villeneuve et al. 2012).

Particulate matter (PM) is a solid or liquid phase object transported in the air. The most common measurement today for particulate matter is PM_{10}, which assesses the mass concentration of particulates smaller than 10 microns in diameter. In recent years, there has been a shift toward measuring $PM_{2.5}$, which is more relevant physiologically (Janssen et al. 2011; Schwartz et al. 2002). While the contribution of PM pollution to premature births was reported by Srám et al. (2005), a review, conducted in 2010, concluded that there is no convincing evidence PM is indeed associated with pre-term birth (Bosetti et al. 2010). A later study reiterated the notion that $PM_{2.5}$ does contribute to risk of pre-term birth (Kloog et al. 2012). The associations between PM levels and hypertension (Dong et al. 2013; Guo et al. 2009; Szyszkowicz et al. 2012), MI (Mustafić et al. 2012) and cerebrovascular diseases (Chan et al. 2006) were established, while no correlation was found between PM and stroke (Villeneuve et al. 2012).

Sulfur dioxide (SO_2) is produced via the burning of sulfur-containing fuel such as oil and coal (Dickey 2000). The associated risk of exposure to higher levels of SO_2 to all the above diseases was well documented (Dong et al. 2013; Guo et al. 2010; Mustafić et al. 2012; Kaplan et al. 2012; Koenig 1999; Srám et al. 2005; Szyszkowicz et al. 2012; Villeneuve et al. 2012).

Volatile organic compounds (VOCs) are a large group of pollutants, including (among others), aromatic hydrocarbons such as benzene, ethylbenzene, m-xylene, p-xylene, o-xylene and toluene (Wallace 1996). VOCs (in particular benzene) have been shown to be strongly correlated to several adverse health effects (Adgate and Goldstein 2014; He et al. 2015; Kim et al. 2013). In a cross-sectional US study, elevated exposure to benzene, ethylbenzene and M&P-Xylenes (MPX) has been linked with wheezing attacks (Arif and Shah 2007). However, VOCs is a large family of pollutants with different characteristics and health effects, which makes its measurement difficult. Further, as VOCs are not part of the set of criteria pollutants, they are significantly less monitored.

The above discussion presents a solid correlation between air pollution and adverse health effects. While this general notion is true, the literature shows that pinpointing specific pollutant's health impact is most difficult if not illusive task. Previous studies that aimed at inferring the connections between air pollutants' ambient levels and adverse health effects, analyzed hospital admissions. In those studies, the air pollution was measured either in the vicinity of the hospital or the patients' residential address. However, the above analysis strongly suggests that finding the correlations between air-pollution and health effects and quantifying the associated risk is difficult. Some of the criteria pollutants have no known pathways and pathologies for causing some of the observed health impacts. In these cases, it is assumed that the measured pollutants are not the direct cause for the disease, but a surrogate indicator (Villeneuve et al. 2012). As the direct causing substance is unknown, the association between the real cause and the surrogate indicator is also unknown, making the correlation assessment even more difficult. Another reason for the weak correlations between hospital admissions and air-pollution is that most of the studies considered air pollution at the hospital or at the patients' home address. The pollution levels at these locations do not necessarily represent the true level where the actual emergency took place. To this end, this study assesses the air-quality at the location of *emergency calls* (rather than the hospitals' location or patients' home address). By evaluating emergency calls, rather than hospital admissions, only extreme situations are considered as in these cases not only medical treatment was needed, the situation was critical and Advance Life Support (ALS) medical team was called to the scene and thus the health impact was prominent.

2 Materials and Methods

2.1 Data

The air quality data consisted of half-hourly ambient air measurements that were acquired by standard air-quality monitoring (AQM) stations in Israel for the time period between 2007 and 2012. These measurements included Benzene, Carbon Monoxide (CO), Ethyl Benzene (E.Benz), M&P Xylene (MPX), Nitrogen Dioxide (NO_2), Nitrogen Monoxide (NO), Nitrogen Oxides (NO_x), O.Xylene (O.X), Ozone (O_3), Particulate Matter under 10 microns in diameter (PM_{10}), Particulate Matter under 2.5 microns in diameter ($PM_{2.5}$), Sulfur Dioxide (SO_2) and Toluene. Each monitoring station measured a different set of pollutants and thus the amount of data available on each of them was different. The air pollution data went through a process of quality control in which data that was incorrect (such as negative values) was removed (Yuval and Broday 2013).

Data acquired from two distinct types of AQM stations were used in this study: environmental- and traffic-oriented stations. The former is typically located on rooftops and towers, designed to measure ambient pollution levels. The latter type is designed to measure transportation related pollution. The inlets of these stations are located in proximity to traffic routes (i.e., lower heights than the inlets of the environmental stations) and typically they measure only pollutants that are known to be related to traffic, such as PM, nitrogen oxides and carbon monoxide. While the inclusion of the two station types into one environmental dataset allows to regard larger dynamic range of nitrogen oxides and carbon monoxides levels, as inherently transportation stations measure higher values of these pollutants, it might bias the results, as traffic stations are not available in all cities.

Yearly city-level categorized emergency ambulance calls were recorded by Magen David Adom (MDA), the Israeli branch of the red cross organization and the national medical emergency service, which serves more than 95% of all ambulance calls in Israel. The categories were set by MDA medical staff on site. The categories included in this study are: parturient (giving birth), hypertension (high blood pressure), breathing difficulties, cerebrovascular accident, CVA, (i.e., stroke) and myocardial infarction (heart attack). It is important to note that the health data was not corrected for the socioeconomic nor any other demographic parameter of the different cities. Thus, the initial findings of this study should be considered as general guidelines for a future, thorough, epidemiological studies.

These two databases had an overlap of 35 cities, accounting for more than half the population of Israel totaling 3,921,900 people in 2007 to 4,043,454 in 2012. This population had a record of 817,639 categorized emergencies during the 5-year time period. Some of these cities had more than one monitoring station. For cities in which more than one station measured a particular pollutant, the mean measurement of all stations in the city was used for the data-point. A data-point is any combination of city and year. That is, for each city, there were 5 data-points (2007, 2008, 2009, 2010, 2011 and 2012), treated distinctly. Examples of data-points are Tel-Aviv 2008, Hadera 2007, Tel-Aviv 2007, Jerusalem 2011, etc.

2.2   Methods

A separate analysis was done for acute and chronic exposure. Chronic exposure was assessed by correlating the mean yearly concentration with emergency calls for each combination of pollutant and emergency type. Acute exposure was assessed by defining a half-hourly measurement that was over 2 standard deviations higher than the national mean as extreme. The rate of extreme measurements for each city in each year was tested as an explaining variable to the number of ambulance calls of each type.

The same correlation tests were conducted for both types of exposure. A multiple linear regression model was made for each combination of pollutant and emergency, when adding the existence of a hospital in the city as a second explaining variable. From this analysis the sign of the pollutant slope was extracted, together with a p-value of the t-test for weather that slope is significantly non-zero.

For further analysis, the data was sliced by pollution levels above or below the median, so that each data-point was marked as "high", "low" or "data not available" with regards to each pollutant. The two groups ("high" and "low") were compared using Analysis of Variance (ANOVA) in order to determine the statistical significance of the difference between their means, taking into account the possibly stratifying parameter of having a hospital present in the city. For the same partitioning, an odds ratio was calculated where the group above the mean were considered "exposed" whilst the group below the mean was considered "not exposed". The odds ratio was calculated using the Cochran-Mantel-Haenszel Estimate, when the presence of a hospital was considered as stratifying variable.

In addition, the same calculations of ANOVA and odds ratio were applied for comparing the top and bottom 25% of exposure to each pollutant (both acute and chronic, separately).

A correlation was considered statistically significant if the p-value computed from the F ratio that was produced by the ANOVA test, or the t-value from the t-test, was smaller than 0.05.

3   Results and Discussion

Tables 1 and 2 detail the correlations between the aforementioned pollutants' chronic and acute exposure respectively for which at least one of the three p-values (t-test for linear regression, ANOVA for top vs. bottom 50% and ANOVA for top vs. bottom 25%) were lower than 0.05. Abbreviations: N.S.—Not Significant, OR —Odds Ratio, N—number of data-points with both a measurement for this pollutant and ambulance data, slope and sign of linear regression correlation (+/−).

The correlations observed for **Parturient** emergency cases show negative correlations to all three nitrogen oxide parameters (NO, NO_2 and NO_x), when comparing above and below the median for both acute and chronic exposure. The volatile organic compound toluene showed a statistically significant negative correlation. On the other hand, carbon monoxide displayed a negative correlation only for acute exposure. $PM_{2.5}$ was positively correlated with parturient emergencies when considering chronic exposure. Acute ozone exposure was the only type of correlation that was statistically significant in all three tests (positive correlation for linear, median and 25%).

It is not clear that parturient emergency calls are directly related to pre-term birth or other adverse effects on the baby. Additional confounding factors should be tested in further research. If we assume parturient ambulance calls are a good indicator of pre-term birth, then there is some agreement and some disagreement

Table 1 Health impact of chronic exposure

Health case	Pollutant	N	Chronic					
			Linear		Top versus bottom 50%		Top versus bottom 25%	
			P-value	Slope	P-value	OR	P-Value	OR
Parturient emergencies	CO	75	N.S.		N.S.		N.S.	
	NO₂	173	N.S.		8.05E-03	0.77	N.S.	
	NO	173	N.S.		8.87E-03	0.77	N.S.	
	NOₓ	173	N.S.		7.80E-04	0.59	N.S.	
	O₃	125	N.S.		N.S.		N.S.	
	PM₂.₅	82	4.22E-02	+	1.99E-03	1.7	N.S.	N.S.
	Toluene	22	6.06E-03	–	8.37E-03	0.41	N.S.	
Hypertension emergencies	NO₂	146	8.60E-03	–	2.42E-02	0.76	N.S.	N.S.
	NOₓ	146	1.75E-02	–	4.96E-02	0.88	N.S.	N.S.
	O₃	105	N.S.		2.34E-02	0.76	N.S.	
Breathing emergencies	Benzene	21	5.10E-04	+	5.00E-02	1.44	8.61E-04	1.7
	CO	75	N.S.		N.S.		N.S.	
	O₃	125	N.S.		2.34E-02	0.76	N.S.	N.S.
Cerebrovascular	Benzene	21	N.S.		N.S.		N.S.	
	CO	75	6.28E-03	–	6.71E-03	0.8	8.00E-03	0.63
	NO₂	125	2.67E-02	–	4.99E-02	0.83	N.S.	N.S.
	PM₂.₅	90	7.82E-04	+	N.S.		6.93E-04	1.02
Heart attacks	NO₂	173	6.39E-03	–	4.48E-03	0.84	4.12E-02	0.77
	PM₁₀	90	N.S.		N.S.		N.S.	

with the data produced here and previous research. Nitrogen oxides and Carbon Monoxide have not been previously linked with pre-term birth by themselves. In this project, a negative correlation emerged. The literature is inconsistent regarding the relationship between PM and pre-term birth. Some claim there is no significant correlation and some claim a positive correlation. In this study, a positive correlation was observed, yet only for $PM_{2.5}$. Previous studies on the relationship between pre-term birth and organic air pollutants were not found. In this project, exposure to toluene was inversely correlated to parturient emergencies.

Hypertension was negatively correlated with chronic exposure to NO_2 and NO_x. Ozone was negatively correlated in one test only with hypertension (median test for chronic exposure). Acute expose for none of the pollutants correlated with hypertension. Except for the Danish cohort (Sørensen et al. 2012), all reviewed literature showed positive links between air pollutants (including NO_2) and hypertension. It should be noted that the Danish cohort was the only study that examined NO_x rather than NO_2. In the analysis presented here, some nitrogen oxides measurements (NO_2 and NO_x) had a negative correlation with high blood pressure. This fits with the Danish cohort (Sørensen et al. 2012). The negative

Table 2 Health impact of acute exposure

Health case	Pollutant	N	Acute					
			Linear		Top versus bottom 50%		Top versus bottom 25%	
			P-value	Slope	P-value	OR	P-Value	OR
Parturient emergencies	CO	75	N.S.		1.14E-02	0.62	N.S.	N.S.
	NO_2	173	N.S.		6.11E-03	0.64	N.S.	N.S.
	NO	173	N.S.		4.54E-03	0.69	N.S.	N.S.
	NO_x	173	N.S.		9.91E-04	0.6	N.S.	N.S.
	O_3	125	2.20E-02	+	8.15E-03	2.31	5.35E-03	2.87
	$PM_{2.5}$	82	N.S.		N.S.		N.S.	N.S.
	Toluene	22	1.78E-02	–	4.03E-02	0.42	N.S.	
Hypertension emergencies	NO_2	146	N.S.		N.S.		N.S.	N.S.
	NO_x	146	N.S.		N.S.		N.S.	
	O_3	105	N.S.		N.S.		N.S.	
Breathing emergencies	Benzene	21	5.87E-03	+	2.19E-03	1.71	4.97E-03	1.7
	CO	75	N.S.		6.62E-03	1.03	N.S.	N.S.
	O_3	125	3.85E-02	–	N.S.		3.10E-02	0.71
Cerebrovascular	Benzene	21	N.S.		2.97E-02	1.37	N.S.	
	CO	75	N.S.		N.S.		N.S.	
	NO_2	125	N.S.		N.S.		N.S.	N.S.
	$PM_{2.5}$	90	N.S.		N.S.		N.S.	N.S.
Heart attacks	NO_2	173	N.S.		N.S.		N.S.	
	PM_{10}	90	N.S.		3.35E-02	0.81	N.S.	

correlation observed here between O_3 and hypertension does not fit previous findings (Dong et al. 2013).

Breathing emergencies were positively correlated in all six statistical tests with benzene. It was also positively correlated with acute exposure to carbon monoxide in one test. A negative correlation was calculated for ozone in three tests out of the six.

One of the strongest correlations found in this study was between benzene (both chronic and acute) and ambulance calls originating from breathing difficulties. This fits well with the previous observations made by Arif and Shah (2007). Ozone showed a negative correlation to breathing-related ambulance calls. A possible explanation for this is O_3 being a secondary pollutant thus having a lower presence in areas with high levels of primary air pollutants. CO was significantly correlated with these emergencies only for acute exposure.

Populations with a higher than the median rate of acute benzene exposure had a statistically significant greater incidence of **cerebrovascular accidents** (CVA). Chronic carbon monoxide showed a negative correlation to stroke in all 3 tests, while no significant correlation was found with acute exposure. NO_2 showed a negative correlation in two of the chronic tests. $PM_{2.5}$ showed a positive correlation in two of the chronic tests. The aforementioned acute-only negative effect of carbon monoxide is interesting especially when considering CVA. Cerebrovascular accidents were significantly less common in populations exposed to elevated chronic CO levels. No such result was found in previous studies. However, a positive link was found in Taipei (Chan et al. 2006). A possible explanation for this may be found in recent medical studies, which found a therapeutic effect for controlled CO release into the body of CVA patients (Bauer et al. 2012). NO_2 also showed negative correlations to CVA, in contradiction to the Taipei study. However, according to Villeneuve et al. (2012), positive correlation to NO_2 may be due to a different traffic-related pollutant that was not measured in Edmonton (i.e., NO2 was served as a surrogate indicator). This could fit with the observed positive correlation with acute Benzene exposure. The positive correlation found for chronic $PM_{2.5}$ and CVA agrees with the Taipei study (Chan et al. 2006) and not with the Edmonton study (which found no significant correlation to PM) (Szyszkowicz et al. 2012).

Myocardial infarctions were more common amongst populations with low exposure to NO_2. A negative correlation was also found for acute PM_{10} exposure. In contradiction to previous data (Mustafić et al. 2012), nitrogen dioxide and PM_{10} displayed a negative correlation with heat attacks. PM_{10}, as discussed previously, is less of an indicator for physiological effect relatively to $PM_{2.5}$, which could explain this discrepancy. With NO_2 the disagreement may be due to a confounding factor not thought of.

4 Conclusions

This study presented an initial evaluation of the correlations and linkage between medical emergency calls and air pollution acute and chronic exposure. The study's findings show that exposure to extreme concentrations of carbon monoxide was correlated with an increased risk for breath-related emergencies, while average ambient concentrations were correlated with a decreased risk of stroke. Both chronic and acute exposure to volatile organic compounds such as benzene were correlated with several types of emergencies, specifically breathing difficulties and cerebrovascular accidents. Acute $PM_{2.5}$ was correlated with an increased rate of childbirth emergency calls.

The study should not be interpreted as giving definitive odds ratios of adverse health effects as a result of exposure to air pollution. The methods used were far too naive for that. With that, this study offers new insight and facilitates the inclusion of air-quality data in medical emergency teams staffing. The research also presents another angle on the difficult analysis of the association between air quality and health. Hence, the new findings presented here should be the basis for a direction of further, more rigorous, research in this field.

Volatile organic compounds (in particular benzene) have been shown to be strongly correlated to several adverse health effects while being the least monitored pollutants. In order to really understand the detriment to public health possibly caused by them, they must be better monitored. It is highly unlikely for any substantial discovery to be made with the current level of VOC data gathering.

References

Adgate J, Goldstein B (2014) Potential public health hazards, exposures and health effects from unconventional natural gas development. Environ Sci

Arif AA, Shah SM (2007) Association between personal exposure to volatile organic compounds and asthma among US adult population. Int Arch Occup Environ Health 80:711–719. doi:10.1007/s00420-007-0183-2

Badr O, Probert S (1994) Sources of atmospheric carbon monoxide. Appl Energy

Badr O, Probert SD (1993) Oxides of nitrogen in the Earth's atmosphere: trends, sources, sinks and environmental impacts. Appl Energy 46:1–67

Bauer M, Witte OW, Heinemann SH (2012) Carbon monoxide and outcome of stroke—a dream CORM true?*. Crit Care Med 40:687–688. doi:10.1097/CCM.0b013e318232d31a

Beelen R, Raaschou-Nielsen O, Stafoggia M (2014) Effects of long-term exposure to air pollution on natural-cause mortality: an analysis of 22 European cohorts within the multicentre ESCAPE project. The Lancet 383:785–795

Bosetti C, Nieuwenhuijsen MJ, Gallus S, Cipriani S, La Vecchia C, Parazzini F (2010) Ambient particulate matter and preterm birth or birth weight: a review of the literature. Arch Toxicol 84:447–460. doi:10.1007/s00204-010-0514-z

Brauer M, Lencar C, Tamburic L, Koehoorn M, Demers P, Karr C (2008) A cohort study of traffic-related air pollution impacts on birth outcomes. Environ Health Perspect 116:680–686. doi:10.1289/ehp.10952

Chan C-C, Chuang K-J, Chien L-C, Chen W-J, Chang W-T (2006) Urban air pollution and emergency admissions for cerebrovascular diseases in Taipei, Taiwan. Eur Heart J 27:1238–1244. doi:10.1093/eurheartj/ehi835

Dickey JH (2000) Selected topics related to occupational exposures Part VII. Air pollution: overview of sources and health effects. Dis Mon 46:566–589. doi:10.1016/S0011-5029(00)90024-5

Dong G-H, Qian Z (Min), Xaverius PK, Trevathan E, Maalouf S, Parker J, Yang L, Liu M-M, Wang D, Ren W-H, Ma W, Wang J, Zelicoff A, Fu Q, Simckes M (2013) Association between long-term air pollution and increased blood pressure and hypertension in China novelty and significance. Hypertension 61:578–58

Guo Y, Jia Y, Wang J, Zhao A, Li G, Pan X (2009) Relationship between ambient air pollution and the hospital emergency room visits for hypertension in Beijing, China. Int J Cardiol 137:S126. doi:10.1016/j.ijcard.2009.09.430

Guo Y, Tong S, Li S, Barnett AG, Yu W, Zhang Y, Pan X (2010) Gaseous air pollution and emergency hospital visits for hypertension in Beijing, China: a time-stratified case-crossover study. Environ Health 9:57. doi:10.1186/1476-069X-9-57

He Z, Li G, Chen J, Huang Y, An T, Zhang C (2015) Pollution characteristics and health risk assessment of volatile organic compounds emitted from different plastic solid waste recycling workshops. Environ Int

Janssen NAH, Hoek G, Simic-Lawson M, Fischer P, van Bree L, ten Brink H, Keuken M, Atkinson RW, Anderson HR, Brunekreef B, Cassee FR (2011) Black carbon as an additional indicator of the adverse health effects of airborne particles compared with PM10 and PM2.5. Environ Health Perspect 119:1691–1699. doi:10.1289/ehp.1003369

Jenkin ME, Clemitshaw KC (2000) Ozone and other secondary photochemical pollutants: chemical processes governing their formation in the planetary boundary layer. Atmos Environ 34:2499–2527

Kaplan GG, Szyszkowicz M, Fichna J, Rowe BH, Porada E, Vincent R, Madsen K, Ghosh S, Storr M (2012) Non-specific abdominal pain and air pollution: a novel association. PLoS ONE 7:e47669. doi:10.1371/journal.pone.0047669

Kim K, Jahan S, Kabir E, Brown R (2013) A review of airborne polycyclic aromatic hydrocarbons (PAHs) and their human health effects. Environ Int

Kloog I, Melly SJ, Ridgway WL, Coull BA, Schwartz J (2012) Using new satellite based exposure methods to study the association between pregnancy $PM_{2.5}$ exposure, premature birth and birth weight in Massachusetts. Environ Health 11:40. doi:10.1186/1476-069X-11-40

Koenig JQ (1999) Air pollution and asthma. J Allergy Clin Immunol 104:717–722. doi:10.1007/978-94-009-4003-1

Mustafić H, Jabre P, Caussin C, Murad MH, Escolano S, Tafflet M, Périer M-C, Marijon E, Vernerey D, Empana J-P, Jouven X (2012) Main air pollutants and myocardial infarction: a systematic review and meta-analysis. JAMA 307:713–721. doi:10.1001/jama.2012.126

Schwartz J, Laden F, Zanobetti A (2002) The concentration-response relation between PM2.5 and daily deaths. Environ Health Perspect 110:1025–1029. doi:10.1289/ehp.021101025

Shah ASV, Langrish JP, Nair H, Mcallister DA, Hunter AL, Donaldson K, Newby DE, Mills NL (2012) Global association of air pollution and heart failure: a systematic review and meta-analysis. The Lancet 2013:1039–1048

Smith K (1987) Biofuels, air pollution, and health: a global review. Springer, New-York, NY, USA

Sørensen M, Hoffmann B, Hvidberg M, Ketzel M, Jensen SS, Andersen ZJ, Tjønneland A, Overvad K, Raaschou-Nielsen O (2012) Long-term exposure to traffic-related air pollution associated with blood pressure and self-reported hypertension in a Danish Cohort. Environ Health Perspect 120:418–424. doi:10.1289/ehp.1103631

Srám RJ, Binková B, Dejmek J, Bobak M (2005) Ambient air pollution and pregnancy outcomes: a review of the literature. Environ Health Perspect 113:375–382. doi:10.1289/ehp.6362

Szyszkowicz M, Rowe BH, Brook RD (2012) Even low levels of ambient air pollutants are associated with increased emergency department visits for hypertension. Can J Cardiol 28:360–366. doi:10.1016/j.cjca.2011.06.011

Villeneuve PJ, Johnson JY, Pasichnyk D, Lowes J, Kirkland S, Rowe BH (2012) Short-term effects of ambient air pollution on stroke: Who is most vulnerable? Sci Total Environ 430:193–201. doi:10.1016/j.scitotenv.2012.05.002

Wallace L (1996) Environmental exposure to benzene. Environ Health Perspect 104:1129

Yuval Bekhor S, Broday DM (2013) Data-driven nonlinear optimisation of a simple air pollution dispersion model generating high resolution spatiotemporal exposure. Atmos Environ 79:261–270. doi:10.1016/j.atmosenv.2013.06.005

Saltwater Intrusion Forecast of the Pleistocene Aquifer Caused by Groundwater Exploiting in the Nam Dinh Coastal Zone

Trinh Hoai Thu, Nguyen Van Nghia, Tran Thi Thuy Huong, Do Van Thang and Nguyen Thi Hien

Abstract The hydrogeological model is applied and carried out for the Pleistocene confined aquifer systems in the Nam Dinh coastal zone, Vietnam. A 3D groundwater flow and matters spreading numerical model forecast about decreasing of the groundwater level and saltwater intrusion on the coastal areas Hydraulic conductivity K, specific storage μ^*, specific yield μ and effective porosity n_0 and boundary conditions are calibrated in this model. The simulated results are significant with low errors corresponding the available data. Calibrated model is utilized to forecast about decreasing of the groundwater level and movement of saltwater intrusion boundary into freshwater areas. The simulated results are indicated that freshwater areas are intrused proportional to exploiting reserves. According to currently freshwater exploiting recharge per year in 2014, freshwater areas until 2025 would be falled 4.17% than in 2005. Moreover, with currently freshwater exploiting

This paper was completed thank to the support of Vietnam Academy of Science and Technology (VAST) project: "*Investigation and assessment salt-freshwater aquifers of Ca Mau province for serving groundwater resources management*" code VAST.ĐTCB.03/17-18 and project "*Research of saltwater intrusion level in Pleistocene aquifer in Nam Dinh coastal zone due to overexploitation of groundwater*", code VAST06.06/14-15.

T.H. Thu (✉) · T.T.T. Huong · D. Van Thang · N.T. Hien
Institute of Marine Geology and Geophysics, VAST, Yuzhno-Sakhalinsk, Russia
e-mail: hoaithu0609@hotmail.com; ththu@imgg.vast.vn

T.T.T. Huong
e-mail: thuyhuong7th@gmail.com

D. Van Thang
e-mail: t.do-van@student.uw.edu.pl

N.T. Hien
e-mail: hienhoageo@gmail.com

N. Van Nghia
Department of Water Resources Management, Viet Nam's Ministry of Natural Resources and Environment, Ha Noi, Vietnam
e-mail: nvnghia@gmail.com

© Springer International Publishing AG 2018
B. Otjacques et al. (eds.), *From Science to Society*, Progress in IS,
DOI 10.1007/978-3-319-65687-8_8

recharge per year in 2014 for consumption and population growth and pumping-wells along main river setting, freshwater areas are intrused more strongly than the above method, and would be decreased 6.16%.

Keywords Saltwater intrusion · Pleistocene aquifer · Hydrogeological model

1 Introduction

Available water resources, including surface water and groundwater, play an important role in the population growth as well as economy in Vietnam. In fact, the negative effects of climate change, untreated sewage water and industrial waste water on surface water have made it more and more vulnerable with the decreasing quality (Ranjan et al. 2006) leads to the possibility that groundwater will become the major resource for the future water supply of Vietnam. Management of this limited resource is necessary for life, development and environment (Wagner et al. 2011). However, during the last decades, the uncontrolled utilization and increasing exploitation of the finite groundwater resources in Vietnam have resulted into several negative effects, especially in coastal areas. One of the major concerns encountered in coastal aquifer is the induced decreasing of the groundwater level continuously and movement of saltwater into freshwater aquifer caused by groundwater over-pumping, known as saltwater intrusion (Wagner et al. 2011).

The hydrogeological model is a 3D groundwater flow numerical model and movement of contaminated matters in groundwater aquifer (Larabi and Smedt 1997; Anderson et al. 2002; Cartwright et al. 2004; Liua 2007a; Lambrakis 2006; Paster and Dagan 2008; Trinh 2012). In this research, the 3D groundwater flow numerical model, Visual Modflow (Grube et al. 2002; Guiguer and Franz 2004; Elkrall et al. 2008; Lindenmaier et al. 2011; Khomine et al. 2011; Trinh et al. 2014; Nguyen 1996) is carried out to simulate and forecast about decreasing of the groundwater level and saltwater intrusion of the Pleistocene aquifer in the Nam Dinh coastal zone, Vietnam. Calibration of model is established by comparison between the currently groundwater level data in 2005 and groundwater level data in this model. Hydraulic properties after calibration are applied in this model for forecasting about decreasing of the groundwater level and saltwater intrusion according the differently exploiting recharges.

2 Hydrogeology of the Studied Area

The studied area is consisted of 5 confined aquifers and 2 confining beds with weak permeability following as:

- The Upper Holocene unconfined aquifer (qh$_2$): The thickness average is ranging from 2 to 28 m. Petrological characteristics are included sand and clay with sandy belong to Thai Binh Formation. The characteristics of groundwater are unconfined aquifer. TDS (Total Dissolved Solid) is ranging from 0.5–23 g/l.

- The Lower Holocene confining beds with weak permeability-Hai Hung Formation ($Q_2^{1-2}hh_1$): The thickness average is ranging from 3 to 40 m. Petrological characteristics are included mostly clay, silty clay which was formed from marine environment with weak permeability.
- The Lower Holocene confined aquifer (qh_1): The thickness average is ranging from 2 to 25 m. Petrological characteristics are included fine grained sand and clay silty sand. The characteristics of groundwater are a weak confined aquifer with 10–15 m column height in pressure. Brackish water is make up low amount in the north area and the remain area is more than 3 g/l in TDS.
- The Lower Holocene confining beds with weak permeability-Vinh Phuc Formation ($amQ_1^3vp_2$): The thickness average is ranging from 7 to 34 m. Petrological characteristics are included mostly grey-green clay and silty clay with weak permeability.
- The Pleistocene confined aquifer (qp): The Pleistocene confined aquifer is the main objective for researching, which is distributed throughout the study area

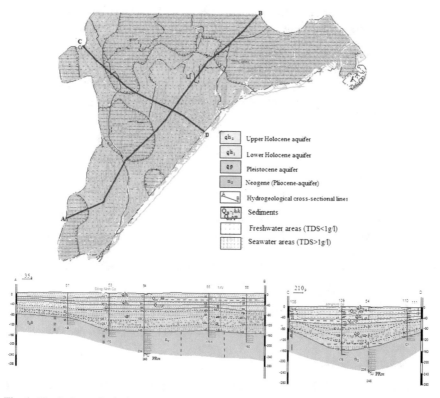

Fig. 1 The hydrogeological cross-section of the Nam Dinh coastal zone and the A-B and C-D hydrogeological cross-sectional lines of the studied area (Doan et al. 2005)

and not exposed on the surface but only appeared in the boreholes. The thickness average is ranging from 10 to 78 m. Petrological characteristics are included sand and gravel belong to the Vinh Phuc Formation, Ha Noi Formation and Vinh Phuc Formation ($Q_1^3 vp$; $Q_1^{1-2} hn$; $Q_1^1 lc$). The characteristics of groundwater are a strong confined aquifer with 50–70 m column height in pressure (Wagner et al. 2011; Anderson et al. 2002; Liua 2007a). Saltwater boundary (TDS = 1 g/l) is indentified clearly by interpolation from boreholes and geophysic data. TDS is more than 1 g/l which is located in the north and east area meanwhile TDS is more than 3 g/l which account to very low area in the northeast. The southwest area has TDS which is smaller than 1 g/l including Hai Hau and Nghia Hung districts which is located on the coastal areas particularly (Doan et al. 2005; Nguyen 2005) (Fig. 1).

3   Materials and Methods

3.1   The Hydrogeological Modelling Method

The principle of salt intrusion forecast method is basing on solution of mathematical flow assignment and mathematical matters spreading assignment:

- Mathematical flow assignment with simulated boundary is spatial distribution of groundwater level height (Guiguer and Franz 2004);
- Mathematical matters spreading assignment with simulated boundary is spatial distribution of saltwater intrusion concentration (TDS > 1 g/l) (Larabi and Smedt 1997; Anderson et al. 2002).

3.2   The Design of Model

The design of grid: The projectional grid of studied area is divided into 14,274 small squares including 122 rows and 117 columns, with equalitical space $\Delta x = \Delta y = 500$ m.

The layers of model: The studied area is consisted of 5 confined aquifers and 2 confining beds with weak permeability.

Hydraulic properties: Hydraulic properties were based on previous hydrogeological studies (Lambrakis 2006; Doan et al. 2005): Hydraulic conductivity K, specific storage μ^*, specific yield μ and effective porosity n_0 which are calibrated in this model.

Boundary conditions:

- River boundary is Day river which has been simulated as the III type relating hydrauly to the Pleistocene confined aquifer (qp).

- Marine boundary is the East Sea which has been simulated as the I type relating hydrauly to the Lower Holocene confined aquifer (qh_1).
- Additional boundary is the west areas bordering confined aquifers and aquifers which were formed from fracture of karst and limestone in Ninh Binh province (Lambrakis 2006).

Additional resources from preparation and evaporite: the Pleistocene confined aquifer (qp) is a deep aquifer about 80–100 m in depth, is covered by the Lower Holocene confining beds with weak permeability-Hai Hung Formation and the Lower Holocene confining beds with weak permeability-Vinh Phuc Formation. Thus, the effect to the Pleistocene confined aquifer is not significant. This research is not carried out with additional resources from preparation and evaporite.

Initial condition:

- Initial condition of mathematical flow is elevational contour map in depth of qh_1 aquifer and qp aquifer in 1/2005 with national monitoring data in the Red River (Nguyen and Trinh 2013)

Table 1 Calibratedhydraulic properties

Aquifers	Zone	K_x (m/ng)	K_y (m/ng)	K_z (m/ng)	μ^*(1/m)	μ	n_0
The upper holocene unconfined aquifer (qh_2)	1	1.5	1.5	0.15	1.6×10^{-7}	0.025	0.025
	2	16	16	1.6	1.8×10^{-6}	0.14	0.14
	3	14	14	1.4	1.6×10^{-6}	0.12	0.12
	4	12	12	1.2	1.6×10^{-6}	0.08	0.08
	5	0.5	0.5	0.05	1×10^{-7}	0.01	0.01
	6	10	10	1	1.6×10^{-6}	0.07	0.07
The Lower Holocene confining beds with weak permeability-Hai Hung Formation ($Q_2^{1-2}hh_1$)	1	0.45	0.45	0.045	2.0×10^{-7}	0.08	0.08
	2	0.25	0.25	0.0025	1.2×10^{-7}	0.03	0.03
	3	0.011	0.011	0.0011	1.0×10^{-7}	0.02	0.02
	4	0.08	0.08	0.008	1.5×10^{-7}	0.05	0.05
The lower holocene confined aquifer (qh_1)	1	12	12	1.2	2.2×10^{-7}	0.18	0.18
	2	8.5	8.5	0.85	2.0×10^{-7}	0.16	0.16
	3	6.5	6.5	0.65	1.8×10^{-7}	0.14	0.14
	4	4.5	4.5	0.45	1.5×10^{-7}	0.12	0.12
	5	1.5	1.5	0.15	1.0×10^{-7}	0.1	0.1
The confining beds with weak permeability-Vinh Phuc Formation ($amQ_1^3vp_2$)	1	0.0006	0.0006	0.00006	1.0×10^{-9}	0.006	0.006
	2	0.0001	0.0001	0.00001	2.0×10^{-9}	0.01	0.01
The Pleistocene confined aquifer qp	1	15	15	1.5	2.0×10^{-8}	0.2	0.2
	2	10	10	1	1.4×10^{-8}	0.14	0.14
	3	6	6	0.6	1.0×10^{-9}	0.12	0.12
	4	20	20	2	1.6×10^{-8}	0.16	0.16
	5	8	8	0.8	1.2×10^{-8}	0.1	0.1

- Initial condition of mathematical movement is TDS spatial distributional map in depth of qp, qh$_1$ and qh$_2$ aquifers (Wagner et al. 2011; Nguyen et al. 2015; Liua 2007b).

Calibration of modelling properties: Probability of model has been carried out by calibration of unstably mathematical inversion assignment and stably mathematical inversion assignment, and comparison with observed values which were computed in national boreholes including Q.108b, Q.109a, Q.110a. If the modelling computed groundwater level and the observed groundwater level have been high error, calibration of input properties was applied. Error after modelling calibration from two assignments lies in the accepted error range (Table 1).

4   The Saltwater Intrusion Forecast Result

According to saltwater intrusion, currently freshwater exploiting recharge per year in 2014 and exploiting and utilizing reserves in the future, the research is designed two methods basing on utilizing and population growth to compute injection of saltwater intrusion into freshwater until 2025.

Scenario 1: Saltwater intrusion forecast for the qp aquifer with currently freshwater exploiting recharge per year in 2014

Currently freshwater exploiting recharge per year (111,630 m^3 per day) and similarly distributed pumping-wells setting for every communes. Calibrated input properties have been applied in model from 2014 to 2017, 2020 and 2025 (Fig. 2):

The saltwater intrusion forecast result for scenario 1:

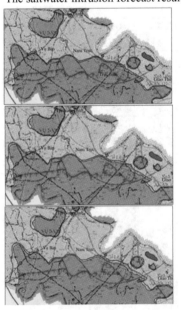

(In 2005) Initial freshwater area is denoted S = 901.64 km^2

In 2017
S = 876.14 km^2 (2.18%)

In 2020
S = 868.29 km^2 (3.79%)

In 2025
S = 864.15 km^2 (4.17%)

Fig. 2 The saltwater intrusion forecast result of qp aquifer basing on method I (the 2017, 2020 and 2025 years)

According to the interpolation result, in the Pleistocene aquifer has appearance of funnel shape which make groundwater level falling with area covering freshwater lenses and the deepest elevation about −12 m (Fig. 5a). Saltwater intrusion has been occurred along interface with small injection. On 10/2025, average movement of interface ranges from 200 to 300 m per year and freshwater areas would be falled 4.17% than in 2005.

Scenario 2: Reasonable arrangement and sustainable freshwater exploiting recharge per year for consumption to ensure decreasing of groundwater level not exceeding permissible limit and reducing saltwater intrusion.

Total exploiting recharge increases over time with levels following as: the 2017, 2020 and 2025 years are respectively 149,531 m^3 per day; 171,198 m^3 per day and 240,115 m^3 per day (Fig. 3).

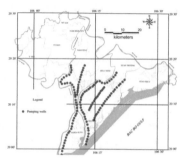

Pumping-wells along main river setting in freshwater areas (Fig.3)

- Pumping-wells exploiting recharge increase over time to consumption and population growth

- Groundwater consumption:

$$Q = P_{population(2014)} * Q_o * (1+r)^n$$

(population growth rate is 0.18% per year)

Fig. 3 Localation of exploiting boreholes according to simulated scenario 2 setting

The saltwater intrusion forecast result for scenario 2:

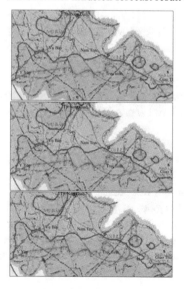

(In 2005) Initial freshwater area is denoted S = 901.64 km^2

In 2017
S = 867.08 km^2(3.79%)

In 2020
S = 849.74 km^2(5.69%)

In 2025
S = 845.44 km^2(6.16%)

Fig. 4 The saltwater intrusion forecast result of qp aquifer basing on scenario 2 (the 2017, 2020 and 2025 years)

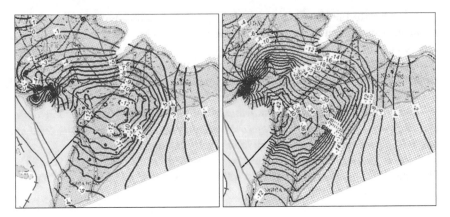

Fig. 5 Elevation of groundwater level forecast of the qp aquifer on October, 2025 (10/2025) basing on scenario 1 (**a**) and scenario 2 (**b**)

Based on reasonable arrangement and sustainable freshwater exploiting recharge per year for consumption, the deepest elevation of groundwater level reducing in the dry season following as the 2017, 2020 and 2025 years are respectively are respectively −26 m, −32 m and −42 m (Fig. 5b). Saltwater intrusion rate is also lied in permissible range about −12 m (Fig. 5a). On 10/2025, freshwater areas would be falled 4.17% than in 2005 with average rate which is about 2.8 km^2 per year (Fig. 4).

5 Conclusions

The hydrogeological model is applied by Visual Modflow software and is carried out for saltwater intrusion forecast of the Pleistocene aquifer in the Nam Dinh coastal zone, Vietnam. Calibrated hydraulic properties and boundary conditions were particularized and standardized by comparison between observed values of groundwater level of the Pleistocene aquifer in the Nam Dinh coastal zone which have been computed from 2005 to 2010 and modelling groundwater level, and is small error.

Calibrated model is utilized to forecast about decreasing of the groundwater level and movement of interface. The simulated results are carried out two exploiting scenarios with differently recharge and setting. Forecast about decreasing of the groundwater level and spreading until 2025 with currently unchanged exploiting recharge per year in 2014 (111,630 m^3 per day) indicate that saltwater intrusion areas are smaller than exploiting recharge relating population growth rate. However, exploiting consumption in a long term is not only at current level but also depending on population growth rate and economic development. Thus, this paper presented method relating population growth rate and pumping-wells along main river setting with exploiting recharge per year for consumption proportionally to population growth rate, ensuring decreasing of groundwater level and saltwater intrusion not exceeding permissible limit and contemporaneously increasing additional resources from rivers for wells which make exploiting recharge of the Pleistocene aquifer higher.

References

Anderson MP, Woessner WW, Hunt RJ (2002) Applied groundwater modeling simulation of flow and advective transport. Academic Press, USA. ISBN-10: 0-12-069485-4

Cartwright N, Li L, Nielsen P (2004) Response of the salt-freshwater interface in a coastal aquifer to a wave-induced groundwater pulse: field observations and modelling. Adv Water Resour 27:297–303

Doan VC, Le TL et al (2005) Groundwater resource of Nam Định Province. J Geol B(25). Ha Noi

Elkrall AB et al (2008) Regioanal groundwater flow modelling of Gash river basin, Sudan. J Appl Sci Environ Sanit 3(3):157–167

Grube A, Hermsdorf A, Lang M, Rechlin B, Steffens A (2002) Numeric and hydrochemical modelling of salt water intrusion into a Pleistocene aquifer—case study Grobbeuthen (Brandenburg). In: 17th Salt Water Intrusion Meeting, Delft, The Netherlands, pp 183–194

Guiguer N, Franz T (2004) Visual modflow. Waterflow Hydrogeologic Sofware, Toronto

Khomine A, János SZ, Balázs K (2011) Potential solutions in prevention of saltwater intrusion: a modelling approach. Adv Res Aquat Environ 1:251–259

Lambrakis N (2006) Multicomponent heterovalent chromatography in aquifers. Modelling salinization and freshening phenomena in field conditons. J Hydrol 323:230–243

Larabi A, Smedt FD (1997) Numerical solution of 3-D groundwater flow involving free boundaries by a fixed finite element method. J Hydrol 201:161–182

Lindenmaier F, Bahls R, Wagner F (2011) Assessment of groundwater resources in Nam Dinh Province, Part B. Improvement groundwater protection Viet Nam

Liua W (2007a) Modeling the influence of river discharge on salwater intrusion and residual circulation in Danshuei River estuary, Taiwan. Cont Shelf Res 27:900–921

Liua W (2007b) Modeling the influence of river discharge on salwater intrusion and residual circulation in Danshuei River estuary, Taiwan. Cont Shelf Res 27:900–921

Nguyen VD (1996) Final report of drawing hydrogeology map Nam Dinh province, scale 1:50.000. General Division of Geology, Ministry of Industry

Nguyen TH (2005) National report of groundwater period 2001–2005. Stored in Department of Geology, Hanoi

Nguyen NT, Trinh HT (2013) Investigation of the saltwater intrusion in the Pleistocene aquifer in the coastal zone of Red River Delta. In: Proceedings of the 11th SEGJ international symposium, Japan

Nguyen TH, Trinh HT, Dinh DC, Vu VM (2015) Study on the freshwater boundary of the Pleistocen aquifer in the coastal zone of Nam Dinh province. In: Proceedings of the 29th EnviroInfo and 3rd ICT4S conference 2015 Copenhagen, Denmark

Paster A, Dagan G (2008) Mixing at the interface between fresh and salt water in 3D steady flow with application to a pumping well in a coastal aquifer. Adv Water Resour 31:1565–1577

Ranjan P, Kazama S, Sawamoto M (2006) Effects of climate change on coastal fresh groundwater resources. Science Direct, no 16. Elsevier, Japan, pp 388–399

Trinh HT (2012) Application of hydrogeological modelling methods in forecasting seawater intrusion of Pleistocene aquifer in Dong Hung and Hung Ha Districts in Thai Binh area. J Mar Sci Technol 12(4A), in Vietnamese. ISSN: 1859-3097, 152-162

Trinh HT, Nguyen NT, Le HM, Vu VM (2014) Application of hydrogeological modelling methods in forecasting seawater intrusion of Pleistocene aquifer in Thai Binh area. In: Proceedings of the 28th EnviroInfo 2014 Conference, Oldenburg, Germany. ISBN: 978-3-8142-2317-9

Wagner F, Trung TD, Phuc HD, Lindenmaier F (2011) Assessment of groundwater resources in Nam Dinh Province. improvement of groundwater protection in Vietnam

Part III
Energy Informatics and Environmental Informatics

Energy Data Management in an Eco Learning Factory with Traditional SME Characteristics

Heiko Thimm

Abstract An Eco Learning Factory is currently being established at an academic institution with the goal to investigate energy management approaches for SME production companies. The article describes a novel conceptual data model and energy data management approaches that are being developed for this factory. The implementation of these approaches is based on the use of an extensible commercial Enterprise Resource Planning (ERP) system and a novel easy to deploy sensor-based energy measurement solution that enables energy consumption measurements for machines without integrated energy meters.

Keywords Energy data management · Energy Management System · Enterprise Resource Planning · Data modelling

1 Introduction

In many manufacturing companies of all industries Energy Management Systems (EnMS) are being established or already used in order to meet supply chain expectations, legal regulations, and also to leverage taxation benefits. In the international standard ISO 50001 (DIN EN ISO 2011) clear requirements for establishing, maintaining, and improving an EnMS are specified. Through an EnMS, companies can obtain transparency about all energy streams in the company. The energy information enables to optimize the company's energy efficiency (BMU 2012).

Due to limited resources, special machinery characteristics, and scarcity of expertise, Small Medium Enterprises (SME) of the production industries sometimes pursue EnMS with a low priority. Often the responsible persons assume that radical changes at all levels (e.g. business processes, IT and industrial automation systems, machines and compounds) are required which result into severe distortions for the

H. Thimm (✉)
School of Engineering, Pforzheim University, Pforzheim, Germany
e-mail: heiko.thimm@hs-pforzheim.de

© Springer International Publishing AG 2018
B. Otjacques et al. (eds.), *From Science to Society*, Progress in IS,
DOI 10.1007/978-3-319-65687-8_9

daily operation and/or over-challenge available capacities. Some companies also seem to question that sufficient benefits for business tasks can be obtained from an EnMS (Meyer 2013). Given this situation, at Pforzheim University we are establishing an Eco Learning Factory that strives to model the typical characteristics of SME production companies concerning the production facilities, the IT infrastructure, and the industrial automation system equipment. In order to prepare the introduction of an intended EnMS, a conceptual data model has been developed for the Eco Factory. Furthermore, a set of complementary energy data management approaches are devised that target the measurement, analysis, and further use of energy data for business tasks. The approach for energy measurement makes use of a sensor-based and easy to install solution that does not impose any specific requirements for the machinery equipment. An extensible ERP system serves as central corner stone for the approach to effectively administer and process the energy data. Through the targeted ERP system extensions, it is possible to improve the energy efficiency and also optimize various general business activities.

In the following Sect. 2, the general characteristics of and the specific role of BIS systems for energy data management in today's SME production companies are described. An overview of the Eco Learning Factory is contained in Sect. 3. Section 4 presents the first research results obtained within the context of the Eco Learning Factory implementation. Related work and concluding remarks are given in Sect. 5 and 6, respectively.

2 Energy Data Management Characteristics of SME Production Companies and the Role of Business Information Systems

EnMS in Today's SME Production Companies. It appears that EnMS can only be found at a minority of today's SME production companies and often these systems are very limited (Meyer 2013). In typical business calculation tasks (e.g. cost accounting, offer calculation, profitability analysis), the firms often work with rough estimates for energy cost rates. The energy consumption of compounds, machines, and devices is usually measured based on only a few centrally installed traditional electric meters. Measurement activities oriented at the real energy consumption of compounds and machines in different operational modes (e.g. standby, standard operation, high load) are only very rarely completed, for example, when new machines are being installed. Typically not all machines are equipped with integrated electric meters and often it is technically not possible or too costly to enhance old machines accordingly. The above described characteristics constitute just some main reasons why in many companies energy management data is not available. There is typically neither data available about the energy demand of productions orders nor about the energy demand of production steps.

Role of Business Information Systems for Energy Management. As of today, there does not yet exist a clear-cut advice for the establishment and deployment of EnMS systems based on the use of a BIS system. More and more new dedicated BIS solutions specialized to EnMS are appearing at the market. But there are also traditional BIS systems available that offer integrated functionalities for energy management. Especially, one can observe a growing number of ERP Systems and MES Systems with integrated energy management functionalities. According to the popular automation pyramid (Vogel-Heuser et al. 2009), in comparison to ERP Systems MES systems are more directly attached to the machinery equipment. Therefore, some experts argue that MES systems are more suited to serve as the basis for EnMS systems. In June 2016, the German VDI (VDI 2016) introduced a corresponding new norm which promotes a MES-based energy management approach. The norm also suggests to extend MES systems by a new module with energy management functionalities.

ERP Systems are for a relatively long time and to a large extend used by SME production companies. They serve the companies as the central and often single data management and transaction processing platform. Therefore, it seems to be a rather natural thought to expect ERP software to already offer integrated functions for energy management. However, in fact only until recently ERP vendors are working on corresponding functional enhancements. Hence, at the present state, one can assume that the EnMS system of only very few SME production companies is based on ERP system functionality.

The standard functionality of ERP systems for production companies usually includes application modules for production planning. These modules typically offer functions for Material Requirements Planning (MRP) and detailed order planning and management. The growing popularity of MES systems has led to the trend that some ERP vendors are integrating typical MES functionalities into their systems. Thus, several available ERP solutions offer functions for operational data and machine data capture and processing, production control stands, functionality for shop floor planning, and further shop floor control functionality (abas software ag 2017).

Dedicated BIS systems for energy management are being deployed by a rising number of firms. Some of these systems make use of business intelligence technologies in order to analyze and forecast energy consumption at different levels (Fusci et al. 2016). Despite the growing popularity of these BIS, one can assume that in the future there will still exist a substantial number of SME production companies that refrain from a deployment of such BIS systems. This decision is often grounded in an "all-in-one" BIS strategy. When changes in the business process landscape occur, companies according to this strategy prefer to adapt and extend their existing ERP system rather than adding a specialized "best of breed" BIS to their IT landscape. Even when there exist gaps within the standard functionality of the ERP system, companies still give priority to customer-specific adaptations and extensions of the ERP system in order to avoid the addition of a new dedicated BIS system.

3 Eco Learning Factory at Pforzheim University

At the German Pforzheim University, we recently launched the Eco Learning Factory. Several research and development labs across different university units are involved in this interdisciplinary research initiative. The Eco Factory is intended to provide an environment for research in the following three areas targeting SME production companies: 1. concepts for the extension of BIS systems towards EnMS-oriented BIS Systems, 2. resource efficiency management, and 3. concepts and methods of the Industry 4.0 vision. As an initial case study for evaluation purposes and also knowledge transfer initiatives, the factory's production system is oriented at the production of bottle openers.

Targeting to mirror the specific characteristics of today's SME production companies (see Sect. 2), it has been decided to use conventional automation technology for the initial architecture of the Eco Learning Factory. It was also decided to make use of an ERP system with integrated MES functionalities. One reason for this decision has been the assumption that ERP Systems most likely offer better support of energy data analyses than MES systems. In fact, there exists a number of ERP systems which offer powerful analytical capabilities. The capabilities are offered on the basis of either directly integrated analytical functionalities or through corresponding add-ons (abas software ag 2017). In contrast to that, MES systems with integrated analytical functionalities are hard to find and often their analytical capabilities are rather limited in comparison to ERP systems.

For the Eco Factory, an ERP system with integrated MES functionalities has been defined as the central platform to manage all the data and to complete most of the operational processes. We opted for the commercial ERP solution of the company abas software, even though the current abas ERP release does not offer integrated energy management functionalities (abas software ag 2017). One of the system's strength is its support of capabilities to adapt and extend the core abas ERP system to company-specific functional requirements. This especially includes capabilities (i) to add a new interface for the import of energy data and (ii) to add new functionalities for energy management. One may easily extend the abas ERP core system by analytical functionalities through a so-called "Business Intelligence" (BI) Plug-In. The BI Plug-In extends the standard abas ERP system by an extensible set of analytical methods to perform complex statistical analyses and data mining analyses on data stored in the abas ERP database.

Specific EnMS Research Goals. As stated above the Eco Learning Factory serves as vehicle for research on future EnMS-oriented BIS systems. In the initial phase, this research objective is focused on electrical energy. It is especially focused at obtaining transparency about the energy consumption of machines and compounds on which production steps are completed. It is one of our central goals to develop and evaluate approaches that enable to obtain transparency of energy demands at production order level and production step level. Obviously, this goal demands more than just simple calculations based on energy data given my machine manufacturer in specification sheets and handbooks. We rather target to obtain precise data (especially considering the actual state of deterioration of the machines) about the machine specific energy consumption for every individual production step. It is intended to achieve this goal by analyses of data that can only made available by frequent energy measurements. The further energy consumers involved in production activities such as e.g. electrical transportation devices, the lightening system and ventilation system but also the automation system and IT system infrastructure are ignored so far.

The particular kind of EnMS that is targeted for the Eco Factory is sometimes referred as "active EnMS" (Franz, et al. 2017). In principle, this means that a continuous improvement process is followed by continuously collecting and analyzing data about energy streams. The analyses (e.g. correlation analyses on data about consumed energy) are oriented at obtaining information to continuously improve the use of energy, the consumption of energy, and the energy efficiency (VDI 2016).

In the Eco Learning factory, we also strive to explore analysis approaches to obtain energy information that is useful for typical business processes of production companies. Our investigations address approaches to extend classical ERP system functionalities by making use of detailed energy data. For example, this includes the consideration of the energy cost in sales order management, pricing, invoices, profitability analyses, cost accounting, quality management processes, and also shop floor scheduling (Thimm et al. 2017).

4 Energy Data Management—Conceptual Data Model and Implementation Approach

It appears that the implementation and maintenance of an effective EnMS presents a number of challenges to companies in general, but especially to SME companies due to budget constraints and limited data management expertise. Given this situation, in the following we describe insights into our approach to establish an effective EnMS for the Eco Learning Factory.

4.1 Conceptual Data Model

Conceptual data models, in general, model the physical and abstract phenomena that are relevant for the given "mini world"—in our case corporate energy management—and their semantic relations. A concept diagram of the model given in the Martin Notation (Martin 1990) is shown in Fig. 1. The boxes denote real world phenomena and the labelled edges model semantic relations between them. The ellipses specify properties of the phenomena. Note that the model contains properties for which we assume that they are important to understand the general idea of

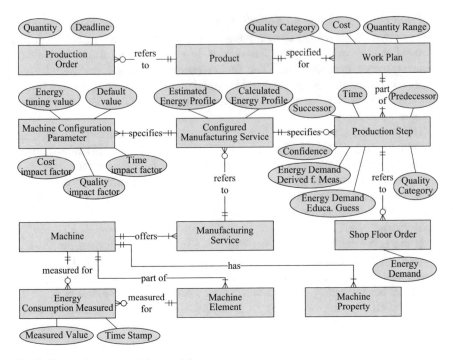

Fig. 1 Proposed conceptual data model

the modelled phenomena. In order to not overload the graphical illustration, the obvious characterizing and identifying properties are not included in the model of Fig. 1.

For a first general understanding consider that *production orders* with assigned *quantities* and *deadlines* are split into a set of *production steps*. This decomposition is defined in the *work plans* of the *products*. Some of the resulting *production steps* require *machines*. The relevant *machine* characteristics are modelled through the concept of *machine property*. In addition to modelling *machines* as a whole, also *machine elements* are modelled. When *production steps* are assigned to *machines,* one needs to consider a set of scheduling constraints. These constraints include the *deadlines* of the orders, the *predecessor/successor* dependencies between the steps of the orders, and the availability of the *machines*.

Now, the model is refined and extended by further concepts that are oriented at collecting, analyzing, and using energy data for energy efficiency optimization. It is a cornerstone of our approach to consider that often there exist several options to complete a *production step* involving different *machines* with different energy demands. For example, it is possible to obtain the raw shape of a metallic bottle opener through the use of a milling machine or a stamp machine or a laser cutting machine. Therefore, the relevant capabilities of the *machines* are generally described in the form of *manufacturing services*. Additionally, *configured manufacturing services* are considered. These services are specialized to the specific requirements of *production steps* through the specification of correspondingly *configured machine parameters*.

The choice of *production steps* that are based on different technologies and usually different *machines*, too, allows to define several alternative *work plans*. The *work plans* typically differ from each other with respect to the *quality* category of the product, the incurred *cost* per product, and the fitting *quantity range*. The details being defined for every *production step* include information about the completion *time*, the energy demand, the *quality category*, and the *predecessor/successor* steps. For the energy demand, a dual attribute approach is used. That is, production steps are associated with two attributes that characterize the respective machine service's energy demand. The first attribute — *Energy Demand Educated Guess*—specifies the energy demand as a result of a well-educated guess considering the energy profiles of the respective *configured manufacturing service*. The second energy data attribute — *Energy Demand Derived from Measurement*—quantifies the energy demand as an absolute number based on respective energy measurement data at machine level. It is possible to use data analysis operations in order to obtain a good approximation value of the production steps' real energy demand (Madan et al. 2013) (Sauer 2016). The absolute value is obtained from further data items addressed in the data model. This includes historic monitoring data about the measured energy demand of completed shop floor orders. The attribute *confidence* signifies the assumed validity of the approximation value. The *configured manufacturing services* are augmented by the *estimated energy profile* and the *calculated energy profile*. The latter profile is calculated from measurement data and, thus, more precise than the *estimated profile*.

The *energy demand* of the *production steps* and the *shop floor orders* are addressed. It is possible to calculate both from the above *energy profiles* and the data of the *production orders* (e.g. quantity). Measurement data is also used to quantify the *energy consumption* of *machines* and the corresponding *machine elements*. Every item of measurement data consists of the *measured value* and a respective *time stamp*. Every *configured manufacturing service* is assigned a set of corresponding *machine configuration parameters*. For every parameter, five different data items are maintained as follows: (i) the *default parameter value* that is used for the *configured manufacturing service*, (ii) a substitution value for the default — referred by *energy tuning value*—in order to tune the energy demand which may possibly impact the cost, the order fulfillment duration, and/or the product quality, (iii) a set of three factors referred by *cost impact, time impact*, and *quality impact* that enable to assess the impact on cost, time, and quality when the *energy tuning value* is chosen.

4.2 Implementation Approach

Similar to many SME production companies, in the ECO Factory we had to cope with the problem that none of our machines is equipped with an integrated electric meter and corresponding enhancement packages from the manufacturers are either not available or too costly. After comprehensive investigations of alternative third party energy metering solutions, we opted for the sensor-based solution of Panoramic Power (Panoramic Power 2016). The sensor equipped clips of the Panoramic Power solution are directly plugged to the main energy supply cables of the machines. The thus measured energy data are frequently collected from the set of installed sensors by a central data collection and analysis station which is managed by the PowerRadar Software. The PowerRadar Unit is able to frequently export energy data in both, the raw measured format and the result format of certain analyses to other systems. On the basis of this approach for measuring energy consumptions of machines and the earlier described abas ERP system (abas software ag 2017) that is used as single central data management platform, we conceptualized the implementation approach shown in Fig. 2.

From the PowerRadar unit the energy consumption data are frequently transmitted ($\sim$ 2 min sampling rate) to the abas ERP system. Note, that these energy consumption data neither include references to production orders nor production steps. On the basis of the Panoramic Power export interface, the abas ERP system receives the measurement data and stores these data in the database. Through standard interfaces and standard functionality, the abas ERP system also obtains shop floor data captured at corresponding data entry units at the machines. For example, this includes data about begin, pause, resume, and termination times of shop floor orders performed by specific manufacturing services. The shop floor data, the EnMS-enhanced master data, and the transaction data (e.g. production orders) are stored in the ERP database, too. Note that by "EnMS-enhanced master data", we mean the data concepts defined in the above proposed conceptual data

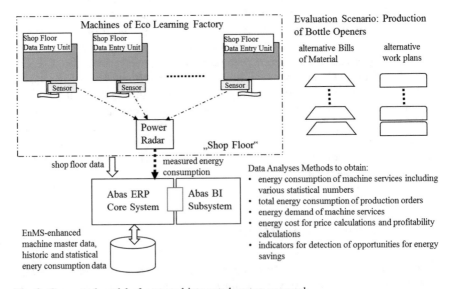

Fig. 2 Conceptual model of proposed integrated system approach

model. This includes machine master data that are extended by energy demand specifications, log data about the energy consumption of production orders, and statistical data about energy consumption at various levels (e.g. machines, production orders, production steps).

The data are obtained by corresponding analyses of data stored in the abas ERP database. By the use of the earlier described abas BI Plug-In, it is possible to perform the analyses on a proven analytical environment that offers an extensible set of ready to use analysis and data mining methods. The list of data analyses methods in the right lower area of Fig. 2 contains examples of methods that are developed on the basis of the abas analytical environment.

5 Related Work

EnMS systems in general and especially in production companies have been addressed by other research groups before (BMU 2012). A reference model to evaluate software solutions for energy efficiency with respect to their fitness for SME companies is being developed in the ReMo Green project (Meyer 2013).The focus of many of the available articles is on (assumed) benefits and general measures to establish an EnMS. The specific aspects of the implementation, maintenance, and use of an EnMS for improving the energy efficiency are described in only a few articles (Langer et al. 2014), however.

Empirical results of energy measurement investigations are described in (Madan et al. 2013). The study also includes investigations of the influence of machine configuration parameters on the load profile of machines. These results are currently under investigation since they seem to be applicable to compute the energy-related machine configuration parameters considered in our approach. Mitsubishi Electric's factory automation energy solution is presented in (Makita et al. 2012). Similar to our approach, the Mitsubishi solution builds on the use of sensors and special energy measuring modules in order to obtain energy data for further use in (business) information systems including ERP systems. It is argued that the use of a MES System interface allows to easily establish an interconnection between the shop floor and the information systems. Instead of the use of a dedicated MES system, in the Eco Factory, the ERP system abas ERP is used which offers integrated MES functionalities.

A general approach to evaluate the "greenness" of ERP systems has been proposed by a research group of the University of Oldenburg (Boltena et al. 2014). The approach is based on the use of an extended set of selection criteria for ERP solutions. The popular set of ERP selection criteria are complemented by further criteria that address environmental friendliness. According to this extended set of criteria, the "greenness" of the abas ERP system will be improved through the intended extension by energy management functionalities.

6   Concluding Remarks

The proposed data model not only contains concepts to integrate typical ERP-based business processing tasks with energy management issues in order to provide transparency about energy consumption at various levels of production companies. The model also constitutes a framework to enhance production scheduling methods so that it is possible to address energy consumption similar to other production scheduling and optimization criteria. We are currently studying a corresponding novel heuristic scheduling method that considers the energy demand of the factory as a further optimization criterion in addition to on-time order fulfillment and low stock levels (Thimm et al. 2017). For the forthcoming implementation of this method, we expect great benefits from the specific integration characteristics of the Eco Learning Factory's BIS implementation architecture.

References

abas software ag (2017) abas Business Suite—The total ERP solution for mid-size companies. http://www.abas.hu/sites/all/files/brossure/abas-erp-production_eng.pdf. Accessed 23 Jan 2017

BMU (2012) Energiemanagementsysteme in der Praxis, ISO 50001: Leitfaden für Unternehmen u. Organisationen. Bundesmin. f. Umwelt, Naturschutz und Reaktorsichheit (BMU). Umweltbundesamt, Berlin

Boltena A, Rapp B, Solsbach A, Marx Gomez J (2014) Towards green ERP systems: the selection driven perspective. In: Proceedings 28th EnviroInfo 2014 conference, 10–12 Sept 2014

DIN EN ISO (2011) Energiemanagementsysteme—Anforderungen mit Anleitung zur Anwendung (ISO 500001:2011). Deutsches Institut für Normung

Franz E, Erler F, Langer T, Schlegel A, Stoldt J, Richter M, Putz M (2017) Requirements and tasks for active energy mangement systems in automotive industry. In: Proceedings 14th golbal conference on sustainable manufacturing, Stellenbosch, South Africa, pp 175–182. doi:10.1016/j.promfg.2017.02.0

Fusci F, Fischer U, Lonij V, Pompey P, Fiot J-B, Chen B, ... Sinn M (2016) Data management system for energy analytics and its application to forecasting. In: Workshop EnDM' 16, proceedings EDBT/ICDT 2016, Bordeaux, France

Langer T, Schlegel A, Stoldt J, Putz M (2014) A model-based approach to energy-saving manufacturing control strategies. In: Proceedings 21st CIRP conference on Life cycle engineering, pp 123–128. doi:10.1016/j.procir.2014.06.019

Madan J, Mani M, Lyons KW (2013) Characterizing energy consumption of the injection molding process. In: Poroceedings ASME 2013 international manufacturing science and engineering conference. doi:10.1115/MSEC2013-1222

Makita H, Shida Y, Nozue N (2012) Factory energy management system using production information. In: Electric M (ed) Mitsubishi Electric ADVANCE, pp 7–11

Martin J (1990) Information engineering: planning and analysis, book II. Prentice-Hall, Englewood Cliffs, NJ

Meyer A (2013) Referenzmodell für eine branchenorientierte Energieeffizienzsoftware für KMU. In: Marx-Gomez J, Lang C, Wohlgemuth V (ed), IT-gestütztes ressourcen-und energiemanagement, proceedings 5. BUIS-Tage. Springer, pp 11–19

Panoramic Power (2016) Panoramic power. http://www.panpwr.com/sensors-and-communications. Accessed 20 Nov 2016

Sauer J (2016) Scheduling regarding energy efficiency. Tagung KI 2016, PuK-WS: Planen/ Scheduling u. Konfigurieren/Entwerfen, p 2016

Thimm H, Kaymakci CO, Tanik M, Andre R (2017) Energy efficiency enhanced shop floor scheduling—Data model and flexible optimization heuristic. In: Proceedings IEEE international conference industrial informatics, Emden, Germany, accepted conference article

VDI (2016) Richtlinie 5600—manufacturing execution systems/fertigungsmanagementsysteme. Verband Deutscher Ingenieure

Vogel-Heuser B, Kegel G, Bender K, Wucherer K (2009) Global information architecture for industrial automation. atp-edition, Automatisierungstechnische Praxis 51:107–115

Methodology for Optimally Sizing a Green Electric and Thermal Eco-Village

Sasan Rafii-Tabrizi, Jean-Régis Hadji-Minaglou, Frank Scholzen and Florin Capitanescu

Abstract This paper presents an energy system for a future eco-village, situated in Luxembourg's city center, whose thermal and electrical energy needs are covered by renewable resources. Specifically, electrical energy is provided by a biogas driven combined heat and power plant, and photovoltaic panels. Lithium-ion accumulators are used for storing the surplus of electrical energy production. Thermal energy needs are mainly covered by a dual source heat pump which draws environmental energy from an ice tank or solar air absorbers. A heat buffer stores heat produced by the combined heat and power plant, a power to heat module and the heat pump. This work focuses on optimally sizing, in terms of power rating, capacity or volume, the energy system components. To this end, a combinatorial simulation-based optimization approach has been developed using MATLAB. The optimal set-up is determined by minimizing the overall cost of the energy system with additional constraints to be respected. Annual needs in thermal and electrical energy, photovoltaic electricity production and the corresponding weather data are based on real data.

Keywords CHP · Heat pump · Green building · Ice storage · Renewable energy · Smart grid · Thermal and electrical energy storage

S. Rafii-Tabrizi (✉)
Faculty of Science, Technology and Communication, University of Luxembourg, Luxembourg, Luxembourg
e-mail: sasan.rafii@uni.lu

J.-R. Hadji-Minaglou
Campus Kirchberg, Luxembourg, Luxembourg
e-mail: jean-regis.minaglou@uni.lu

F. Scholzen
Luxembourg Institute of Science and Technology, Luxembourg, Luxembourg
e-mail: frank.scholzen@uni.lu

F. Capitanescu
Esch-Sur-Alzette, Luxembourg, Luxembourg
e-mail: florin.capitanescu@list.lu

© Springer International Publishing AG 2018
B. Otjacques et al. (eds.), *From Science to Society*, Progress in IS,
DOI 10.1007/978-3-319-65687-8_10

1   Introduction

The European Union endorses the mandatory target that a share of 20% from the Europe-wide energy consumption is produced from renewable energy sources whereas all Member States have to achieve a minimum of 10% by 2020 (European parliament 2009). The Luxembourgish government has set a clear schedule for increasing the share of renewable energy production on a national level. The national plan foresees that 11% of Luxembourg's gross final energy consumption must be produced by renewable energies by the end of 2020 (Gouvernement du Grand-Duché de Luxembourg 2016). To achieve this objective, the Luxembourgish government has set out three areas (Gouvernement du Grand-Duché de Luxembourg 2016):

1. Developing renewable energy on a national level by means of producing electricity and heating/cooling from renewable resources
2. Adding biofuels to fuels for domestic use and development of electrical mobility in the private and public sector
3. Application of cooperation mechanisms realized through statistical transfers and joint projects with other Member States of the EU and third countries.

A description of these measures as well as a detailed timeline considering their implementation can be found in Luxembourg's action plan for renewable energies (Gouvernement du Grand-Duché de Luxembourg 2008). Point 1 and 2 are being theoretically and practically addressed in a project which is realized in Luxembourg's city center. An eco-village is being planned by a family owned Luxembourgish construction company. The location is a 3.6-hectare ex-industrial site in the heart of Luxembourg city.

2   Electrical and Thermal Load Profiles

This eco-village adheres to the Bioregional's[1] zero carbon principle which means not only constructing energy efficient buildings,[2] but also covering thermal and electrical energy needs with renewable resources. A major requirement of the design phase is that 100% of thermal and at least 20% of electrical energy must be produced on site. The remaining electrical energy is produced off site. In this preliminary study, the onsite production of 100% thermal and 100% electrical energy is investigated. The total annual amount of thermal and electrical energy needs is provided by the construction company as shown in Table 1.

---

[1]Bioregional is a NGO which has developed the One Planet Living Principles.
[2]Standard passive AAA following the Luxembourgish legislation.

Table 1 Estimated annual energy needs

Energy type	Energy (kWh/a)
Heating	738200
Hot water	544400
Cooling	186300
Electricity building	849500

Because the project is at an incipient planning phase, detailed architectural plans or thermal and electrical load profiles are not available yet. Electrical and thermal load profiles had to be set up artificially to meet the annual energy requirements. The project foresees not only room for apartments and hotels but also offices, shops and recreational possibilities. Consequently, the electrical load profile of the building has been adjusted by combining H0 and G0 standard load profiles that contain electrical load demands for each quarter of an hour during one year (Stadtwerke Ulm 2016a; Stadtwerke Ulm 2016b) (Fig. 1).

These have been extrapolated to the annual energy requirements of the building. Figure 2 shows an exemplary daily electrical load profile of the building during the simulation. Electrical needs are summed up hourly. The needs in thermal energy for domestic hot water are distributed equally on hourly basis. The needs in thermal energy for heating $Q_{heating}$ and cooling $Q_{cooling}$ are calculated depending on the ambient temperature. For this, two threshold temperatures $T_{cooling}$ and $T_{heating}$ must be defined to indicate above or under which outside temperature cooling or heating is required. In a next step the arrays of temperature differences $\Delta T_{cooling}$ and $\Delta T_{heating}$ are calculated. $\Delta T_{cooling}$ and $\Delta T_{heating}$ contain the hourly needs of cooling or heating energy for one year. Depending on the ambient temperature, the elements of the array are equal to 0 or to the difference between the ambient and the corresponding threshold temperature. The arrays $Q_{cooling}$ and $Q_{heating}$ containing the

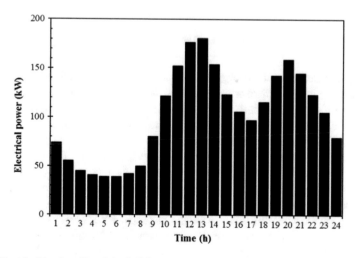

Fig. 1 Electrical load profile of the building

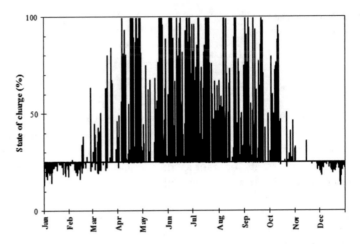

Fig. 2 State of charge of the accumulator

hourly energy needs for cooling and heating are calculated by means of the Manhattan norm according to the Eqs. (1) and (2). The arrays have 8760 rows and 1 column each.

$$Q_{cooling} = \frac{186300}{\left\| \Delta T_{cooling} \right\|_1} * \Delta T_{cooling} \tag{1}$$

$$Q_{heating} = \frac{738200}{\left\| \Delta T_{heating} \right\|_1} * \Delta T_{heating} \tag{2}$$

3   Energy System Components

The proposed design of the energy system can be subdivided in four parts: electrical energy production, electrical energy storage, thermal energy production and thermal energy storage, whose detailed description is provided hereafter. The main renewable electrical energy producing technologies are photovoltaic panels PV and a biogas-fed combined heat and power plant CHPP. PV's are installed on the rooftop of the building whose surface is limited to 9035 m². The CHPP provides the base load of the electrical needs. The PV power output is volatile and not controllable. The CHPP requests maintenance on a regular basis. To prevent power outages and to ensure the electrical power balance, the eco-village is connected to the public power grid utility. To store the electrical PV energy in overproduction periods and use it at times of lack of renewable electricity production, Li-Ion accumulators ACC are chosen to increase the electrical autarky. The CHPP produces a constant amount of electricity. The power output overcomes the electrical needs in the morning. The surplus is stored in the accumulators and can be accessed during peak times at noon and in the evening. A dual source heat pump is installed as a heat producing unit to cover the needs in heating and domestic hot water. During the heating season and thermal energy production for domestic hot water, the heat pump draws environmental energy from an underground ice tank or solar air absorbers. The environmental heat sources are chosen depending on the temperature level in the ice tank and the ambient temperature. The temperature in the ice tank drops until the solidification process of water occurs, releasing its latent energy. The transition from water to ice, requires an amount of energy, which equals the amount of energy used for the inverse process of thawing. During summer, the ice tank is used to cool the building through natural cooling. Natural cooling does not request an operational compressor. Hence the need in electrical energy can be reduced. Giving the fact that more heating than cooling energy is needed, the ice tank reaches low temperatures which decrease the coefficient of performance of the heat pump. A heat source management determines whether environmental energy should be taken from the solar air absorbers or the ice tank. The solar air absorbers and the ice tank surrounding soil contribute to its regeneration. The biogas fed CHPP produces heat as a byproduct. This heat can be used for domestic hot water and heating. Due to financial constraints, the maximal capacity of the accumulators is limited. Once the maximal capacity is reached, the surplus of electrical energy is transformed to heat. As a power to heat module, an electrode boiler is used. In a further study the dual source heat pump is investigated as a power to heat module. Heat produced by the heat pump, the CHPP and the electrode boiler is stored in a sensible heat buffer. The ice tank acts as an environmental heat source of the dual source heat pump. The ice tank delivers environmental energy for the heat pump and is installed underground to recuperate energy from the soil, contributing by this to its regeneration.

4 Mathematical Model

The formulation of the presented energy system requires a mathematical description of the thermal and electrical storage and production elements. This section describes a simplified mathematical model of the eco-village operation in a form of an annual simulation of the combined electrical and thermal behavior of the eco-village, where the following aspects have been neglected: transmission losses, reactive power balance, electric circuit loading, voltage profile, electrical energy consumption of plant components such as pumps, valves or sensors, thermal losses of the heat storage and piping. The operation of storage devices is pre-determined and described as follows. The state of charge of the accumulator SOC is determined depending on the electrical balance of the energy system. The electrical balance EB represents the difference between electrical energy production and electrical energy consumption. The hourly self-discharge rate HSDR, the deload efficiency η_{acc}, the depth of discharge DoD and maximal capacity C are parameters of the accumulator. The state of charge of the accumulator is calculated as follows:

$$\mathbf{SOC(t)} = \begin{cases} \min\left((1 - HSDR) * \mathbf{SOC(t-1)} + \mathbf{EB(t)}, C\right) & \text{if } \mathbf{EB(t)} \geq 0 \\ \max\left((1 - HSDR) * \mathbf{SOC(t-1)} + \frac{\mathbf{EB(t)}}{\eta_{acc}}, (1 - \text{DoD}) * C\right) & \text{if } \mathbf{EB(t)} < 0 \end{cases}$$

$$(3)$$

The CHPP charges the accumulator as long as the electrical energy production exceeds the consumption of the site and the storage limit of the accumulator is not reached. The plant CHPP operates at full load until the accumulator's state of charge equals the annual peak load APL of the eco-village at any time of the year:

$$\mathbf{CHPP_{elec}(t)} = \min\left(\mathbf{CHPP_{elec,max}}, \ \max\left(\frac{APL}{\eta_{acc}} + (1 - DoD) * C - (1 - HSDR)\right.\right.$$
$$\left.\left. * \mathbf{SOC(t-1)}, 0\right)\right)$$

$$(4)$$

The heat pump HP keeps the heat storage HST at a constant temperature of 45 °C. The temperature inside of the heat storage varies due to thermal demands TD for heating and domestic hot water. The heat pump stops working once the temperature inside the heat buffer exceeds 45 °C. Depending on the temperature levels, the heat pump draws environmental energy from the ice tank IST or the solar air absorbers SAABS. The heat pump coefficient of performance is calculated dynamically on the basis of Carnot's efficiency multiplied by the degree of quality of the heat pump. The thermal energy production of the heat pump is calculated in the Eq. 5:

$$\mathbf{HP_{th}(t)} = \max\left(\mathbf{TD(t)} - \mathbf{HST(t)}, 0\right) \tag{5}$$

The upper limit temperature inside the heat buffer is 95 °C. The heat buffer is charged through the thermal output of the CHPP, the power to heat module and the heat pump. The discharging process is realized through thermal needs in heating and domestic hot water. No heat loss with the environment is considered. The mathematical description of the ice tank requires information on temperature, mass of ice and mass of water of the prior time step. The thermal energy balance of the ice tank considers heat exchange with the soil, thermal energy drawn by the heat pump and thermal energy added through natural cooling. The temperature of the soil is supposed to be constant at 10 °C. The heat exchange coefficient with the soil is fixed at 0.015 kWh/m^2 (Eschborn 2016). The ice tank works within a temperature range of −7 to 20 °C. On basis of the temperature, mass of ice and mass of water of the prior time step, the thermal balance enables the calculation of temperature, mass of ice and mass of water of the recent time step. Depending on the thermal energy balance, a single or a combination of the basic equations for latent or sensible heat transfer are used to determine the temperature, mass of ice and mass of water of the recent time step. The power output of PV is assumed to take the values of real data which are normalized with the power rating. The optimization process requires information on the total cost of the energy system. The total proportionate cost for a fixed interval of one year TC of the energy system is composed of the cost of investment CoI, the fix operating cost FOC, the variable operating cost VOC and the fuel cost FC of each element (Wirth, Harry 2016; Institut Für Stromrichtertechnik Und Elektrische Antriebe 2012; Viessmann: Login Professionel 2016; Raab 2006; Ellen and Berger 2015; Recknagel et al. 2009). The investigated period of time is 20 years.

$$
\begin{aligned}
TC = &CoI_{PV} * PV_P + FOC_{PV} * PV_P + CoI_{elec,max} * CHPP_{elec,max} + VOC_{CHPP} * \|\mathbf{CHPP_{elec}}\|_1 \\
&+ FC * \frac{\|\mathbf{CHPP_{elec}}\|_1}{\eta_{elec}} + CoI_{HP} * HP_{th,max} + CoI_{ACC,C} * C + FOC_{ACC} * C \\
&+ CoI_{ACC,P} * ACC_{P,max} + CoI_{HST} * HST_V + CoI * SAABS_{P,max} + CoI_{IST} * IST_V \\
&+ CoI_{P2H} * P2H_{P,max} + CoI_{elec,tpg} * \|\mathbf{Elec_{tpg}}\|_1
\end{aligned}
$$

$$(6)$$

5 Optimization Approach

In the current study, a simulation-based optimization scheme as shown on Fig. 4 is developed to investigate the ideal set-up of the energy system. In this scheme a computer model of the energy system is first created in MATLAB as described in Sect. 4. The optimization variables are the volumes of heat and ice storage, the power ratings of PV and CHPP and the capacity of the accumulators. Thermal and electrical loads are established as described in Sect. 2. The input data includes information on time, temperature, electrical and thermal load distribution of the building, normalized real data on PV power output, power rating of PV and CHPP, capacity of the accumulator, volumes of the heat and ice buffer. To each element of the energy system is attributed a depending array of data of a specific nature containing discrete information concerning the power rating, the capacity or the volume, representing the elements of the array as well. To find the optimal set-up of energy producing and storing technologies, all possible combinations of the vectors are determined. Five arrays containing n elements each would result in n^5 combinations. Depending on the simulation the total number of combinations varies between 27,225 and 320,166. The input file is processed by the MATLAB code, creating output data. The output data are filtered with respect to four constraints (1–4) and one main objective (5) with EXCEL's filter functions:

1. No electricity taken from the public power grid
2. No electricity delivered to the public power grid

3. Enough storage for heat produced by the CHPP
4. Amount of active cooling not exceeding 55% of the actual cooling need
5. Selection of the combination with the lowest cost

An optimization cycle results in a combination, which meets the criteria (1–4) at any time and the main objective (5). Electrical autarky is determined by comparing the charge of the accumulator and the electrical balance at any given time step. Input files for the next optimization cycle are set up by refining the power rate, capacity and volume arrays with focus on the prior results.

6   Simulation and Type of Results

This section provides the results obtained with the proposed optimization methodology. Each simulation is executed for a period of one year based on an hourly time step. The optimal combination of the components must be determined by applying the previously mentioned filtering criteria. A set-up is considered optimal if the percentage difference of the total cost PDTC of the energy system lies beneath at least one percent compared to the previous simulation cycle. Once the optimal set-up is determined, the temporal behavior of each component is analyzed more in detail. Table 2 represents the final set-up after each simulation after application of the selection criteria.

Analyzing the results after each simulation cycle, shows that the total cost TC function is converging to an optimal solution. The prior set one percent criteria applied on the percentage difference of the total cost is achieved after four simulation cycles. The temporal behavior of the final set-up is analyzed for a period of one year. The optimal values of the power rate (P), capacity (C) or volume (V) of each component correspond to the results of the fourth simulation cycle in Table 2. PV power outputs vary between 0 and 333 kW. CHPP power outputs vary between 0 and 169 kW. A comparison of the temporal behavior states the fact that the power output of CHPP decreases as the PV production increases. Figures 2 and 3 display the state of charge of the accumulator and the surplus of electricity which is transformed to heat. The CHPP charges the accumulators until the state of charge reaches 25%, which is equivalent to 316 kWh. Further charging is realized through the power output of PV during which the power output of CHPP equals zero. If the

Table 2 Results of the simulation

Simulation	P_{pv} (kWp)	$P_{elec,chpp}$ (kW)	C_{acc} (kWh)	V_{hst} (m^3)	V_{ist} (m^3)	TC (€/a)	PDTC (%)
1	474	198	1200	1142	950	361979.4	–
2	474	176	1250	176	760	327519	9.52
3	415	170	1260	176	779	322049.4	1.67
4	413	169	1265	184	779	321598.5	0.14

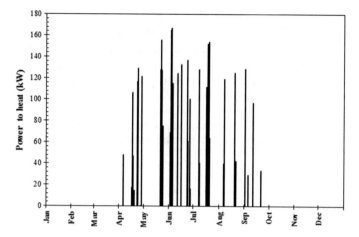

Fig. 3 Surplus electricity transformed to heat

state of charge of the accumulator reaches 100%, any surplus of electricity is transformed to heat and stored in the heat buffer. On top of that, high ambient temperatures during the month of November reduce the thermal needs for heating. The temperature inside the ice tank ranges from −2 to 20 °C. A minimum occurs during March. Maximal temperature values are reached during the months of July to September. The reason for this increase of temperature is additional thermal input through natural cooling. From November to July thermal energy is drawn from the surrounding soil because temperature inside the ice tank drops below 10 °C. During the month of November, the heat pump draws environmental energy from the solar air absorbers. Hence the temperature inside the ice tank increases due to heat recovered from the soil.

7 Conclusion and Future Work

A study of a proposed energy system for a future eco-village situated in Luxembourg city has been carried out. The energy system consists of thermal and electrical energy production and storage components. This paper describes a methodology to find the optimal size of various components to achieve 100% of electrical energy production on site from renewable energy sources. A simulator has been developed to mimic the temporal behavior of these components throughout a year. The simulation results are used to find the optimal volumes, power ratings and capacities of thermal and electrical energy storage and production devices. This problem has to satisfy multiple predefined selection criteria. The selection criteria are applied to filter the simulation results for the ideal set-up, leading to an optimal solution after four simulation cycles which takes about 10 h of computational time in total. The proposed paper represents the preliminary work for an experimental laboratory set-up which will be constructed with reduced scale.

Further studies will be carried out to determine the potential impact on this deterministic optimal design of future price variations of electricity taken from the public power grid as well as technological parts of the energy system such as photovoltaic panels and accumulators.

Acknowledgements The presented work is framed in a PhD project at the University of Luxembourg within the Research Unit in Engineering Sciences. We thank Jörg Petermann for the prior work he provided for this project.

References

Bundesamt für Wirtschaft und Ausfuhrkontrolle: Merkblatt Wärme-und Kältespeicher. Eschborn (2016)

Ellen D, Berger W (2015) Integration Erneuerbarer Energien Mit Power-to-Heat in Deutschland.: Springer Vieweg, Wiesbaden

European parliament: directive 2009/28/EC of the european parliament and of the council of 23 (2009) http://eur-lex.europa.eu/legal-content/EN/TXT/PDF/?uri=CELEX:32009L0028&from=DE. Accessed 10 Dec 2016

Gouvernement du Grand-Duché de Luxembourg: Luxembourg action plan for renewable energy (2008) https://ec.europa.eu/energy/sites/ener/files/documents/dir_2009_0028_action_plan_luxembourg.zip. Accessed 6 Dec 2016

Gouvernement du Grand-Duché de Luxembourg: national plan for smart sustainable and inclusive growth (2016) http://ec.europa.eu/europe2020/pdf/csr2016/nrp2016_luxembourg_en.pdf. Accessed 8 Dec 2016

Institut Für Stromrichtertechnik Und Elektrische Antriebe: Technologischer Überblick Zur Speicherung Von Elektrizität. Aachen (2012)

Raab S (2006) Simulation, Wirtschaftlichkeit Und Auslegung Solar Unterstützter Nahwärmesysteme Mit Heißwasser-Wärmespeicher. Cuvillier, Göttingen

Recknagel H, Sprenger E, Schramek E-R (2009) Taschenbuch Für Heizung Und Klimatechnik. Oldenbourg Industrieverlag, München

Stadtwerke Ulm: Ulm Netze (2016a) https://www.ulm-netze.de/fileadmin/content/downloadcenter/Strom/Lastprofile/Netze-Strom-Lastprofil-SH0_2016.csv. Accessed 12 July 2016

Stadtwerke Ulm: Ulm Netze (2016b) https://www.ulm-netze.de/fileadmin/content/downloadcenter/Strom/Lastprofile/Netze-Strom-Lastprofil-SG0_2016.csv. Accessed 12 July 2016

Viessmann: Login Professionel (2016) http://www.viessmann.lu/fr/services-viessmann/login.html. Accessed 18 June 2016

Wirth, Harry: Recent Facts about Photovoltaics in Germany. Freiburg (2016)

Multi-Model-Approach Towards Decentralized Corporate Energy Systems

Christine Koppenhoefer, Jan Fauser and Dieter Hertweck

Abstract Decentralized energy systems are characterized by an ad hoc planing. The missing integration of energy objectives into business strategy creates difficulties resulting in inefficient energy architectures and decisions. Practice-proven methods such as Balanced Scorecard, Enterprise Architecture Management and Value Network approach supports the transformation path towards an effective decentralized system. The methods are evaluated based on a case study. Managing multi-dimensionality, high complexity and multiple actors are the main drivers for an effective and efficient energy management system. The underlying basis to gain the positive impacts of these methods on decentralized corporate energy systems is digitization of energy data and processes.

Keywords Transformation path · Decentralized corporate energy system · Balanced scorecard · Enterprise architecture management · Value network

1 Introduction

The number of decentralized power generating sites has increased in recent years. Enterprises discovered decentralized energy generation for different reasons such as price advantages, planning stability, declining amortization time of power facilities, flexibility of energy demand and environment/resource protection (DIHK 2014; Döring 2015). However research shows, that corporate energy management is seldom aligned with the overall corporate strategy ending in poor energy system decisions or in business decisions contrary to energy goals (Manthey and Pietsch

C. Koppenhoefer (✉) · J. Fauser · D. Hertweck
Reutlingen University, Reutlingen, Germany
e-mail: christine.koppenhoefer@reutlingen-university.de

J. Fauser
e-mail: jan.fauser@reutlingen-university.de

D. Hertweck
e-mail: dieter.hertweck@reutlingen-university.de

© Springer International Publishing AG 2018
B. Otjacques et al. (eds.), *From Science to Society*, Progress in IS,
DOI 10.1007/978-3-319-65687-8_11

2013; Posch 2011). This leads to "accidental" energy architectures (Giroti 2009) which result in inefficient energy savings or efficiency measures. The digital transformation in the energy sector is an opportunity, as well as a thread to overcome these barriers (Doleski 2016).

Our paper analyzes the challenges of a digitized transforming path towards corporate decentralized energy systems (CDES). It researches the applicability of best-practice methods to develop effective CDES. The feasibility of the approach using methods like e3 Value (e3 V), Balanced Scorecard (BS) or Enterprise Architecture Management (EAM) and its ability to detect transformation hurdles are demonstrated on a case study.

2 Methodology

The aim of this research is to analyze, design, implement and evaluate a transformation path towards a digitized decentralized energy management system. A case study approach was selected to overcome the current actual gap in empirical research. Case studies are useful to research a new domain or phenomenon, providing initial hypotheses, which may be tested later systematically on a larger number of cases (Flyvbjerg 2006).

Our case is a 300 employee, medium-sized enterprise in the tourist sector that offers a unique botanical garden to its visitors. It has several decentralized renewable energy systems on its property and the ecological goals are clearly stated in its corporate strategy. When the enterprise started implementing an energy management system, various obstacles were detected.

Since the authors adopt an active role in the transformation process, our research activity conforms to the tenets of action research (AR). Baskerville (1999) defines action research as an interactive process involving researchers and practitioners acting together on a particular cycle of activities, including problem diagnosis, action intervention, and reflective learning. Three data collection methods were used during the case study: participating observation, semi-structured interviews and document analysis.

Conducted research is part of a public funded research program named ENsource (www.ensource.de) which focuses on decentralized, flexible solutions for future energy production and distribution. This project is funded by the Ministry of Science, Research and the Arts of the State of Baden-Wuerttemberg, Germany and the European Regional Development Fund (EFRE).

3 Case Analysis

The implementation of renewable energy generators on company´s territory is based on the corporate vision to create an economic and ecological balance. The company built up several energy generation sites in the past years to achieve this corporate vision but also faces high energy demands due to green houses and several catering facilities. Since 2013, the company has started to build up an energy management system.

The scientific literature concerned with energy management systems (EnMS) is still limited and publications focus on practice-oriented books (Hubbuch 2016). A report from Natural Resources Canada proposes a best practice framework based on results of thousands of trainings conducted with organizations in energy management (Natural Resources Canada 2015). According to this guide, effective energy management requires a holistic approach that considers actions in eight categories: Commitment, Planning, Organization, Projects, Financing, Tracking, Communication, and Training. The category Value Network Management has not yet been part of the best-practice energy management. Due to the case study findings and its discussed relevance in scientific literature (Hertweck 2016) the category "Value Network Management" is integrated into the energy management frame-work. The performance in each category is rated on a Likert-scale of 1–5. Level 5 means the organization works in an optimal way, while level 1 means no action can be noted in this category. This framework can be used to set an energy policy or to check the state of an EnMS within an organization.

The findings of the case study are examined and rated using this framework to detect gaps in the company's decentralized corporate energy system and then to identify practice-proven methods aiming to close these gaps.

Figure 1 gives an overview of the final case study ratings. The results show that the company and its management board have a high commitment towards energy objectives. They have started energy transformation measures in different areas. Yet, further progress is hindered through two specific categories: planning and tracking. A third category "multiple actors" is of growing relevance since the

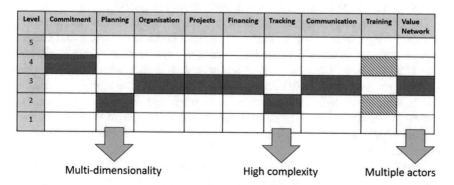

Fig. 1 Overview of the ratings and resulting resolving challenges

achieving of the company´s energy goals relies on its value network. A detailed description of the different dimensions and ratings are published in Koppenhöfer et al. (2017).

The corporate vision of a balanced economical-ecological strategy (criteria "Commitment") requires an alignment of economic and energy-related objectives and measures (criteria "Planning"). The implemented EnMS hasn't supported such functionality so far. Therefore, the reflection of energy items to corporate decisions has not been implemented. The corporate decision making processes are mostly economically driven. To overcome this obstacle **multi-dimensional viewpoints** have to be integrated into the company's strategy.

Managing a decentralized corporate energy system is based on energy data. The category "Tracking" shows an essential gap in this area. Missing measurement spots, heterogeneous data formats and poor digitized processes are the main barriers. On top of that, the high diversity of power systems leads to **highly complexity**.

The case analysis also showed that decentralized corporate energy systems are not based on traditional supplier-buyer-value chains anymore, but encompass various actors that form an ecosystem of interdependent value-added services. This **multiple-actor-approach** is of specific means and has to be included in a management framework.

4   Identification and Transfer of Practice-Proven Methods

Multi methods, reflecting the identified challenges support the transformation path towards an effective CDES (Fig. 2). Integrating different viewpoints into a company´s strategy can be achieved by using the Balanced Scorecard Method (BSC). The Enterprise Architecture Management (EAM) approach has been proven to be an efficient instrument to align business and IT and to control the complexity of IT landscapes (Hanschke 2012; Hauder et al. 2014) and reflects as well the energy domain (Uslar et al. 2013). Decentralized energy systems depend on different actors and form a value network (Christensen and Rosenbloom 1995).

The BSC approach (Kaplan and Norton 1992) addresses a combination of four business perspectives (financial, customer, internal business processes, and learning and growth) and offers the opportunity to integrate further strategic views (Kaplan and Norton 1996). The BSC is a global standard, widely used in private (Van Grembergen and Saull 2001) and public sectors (U.S. Departement of Energy 2017). Objectives are linked to measures and quantified through key performance indicators (KPI). A constant controlling of improvements is possible. Such a holistic approach enables the integration of energy objectives into corporate strategy dimensions.

EAM is used to model the as-is landscape and to derive from there the to-be state. The decentralized energy system reflects a complex energy architecture, which has to be aligned to the IS and enterprise architecture. Therefore, EAM

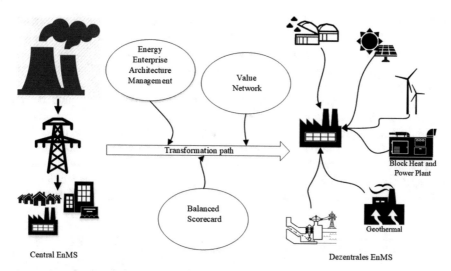

Fig. 2 Transformation path

methods provide a positive impact for establishing an effective EnMS. They have
been accepted in academia and practice (Uslar et al. 2013) for years.

Today enterprises can seldom be viewed as isolated entities. Usually they are
connected in an enterprise network forming an ecosystem (Krcmar 2011). Many
enterprises have established energy value networks with cooperating partners.
Therefore, it is necessary to model and integrate the value flow into the EnMS of
each partner.

4.1 Improving EnMS with the Balanced Scorecard Method

The BSC method offers several advantages: (i) it enables the monitoring of present
performances as well as obtaining information about the future ability to perform.
(ii) It assists in translating an organization's vision into actions through strategic
objectives and a set of performance measures supported by specific targets and
concrete initiatives. (iii) Using the BSC facilitates the identification of success
drivers, allowing managers to focus on a small number of critical indicators, thereby
avoiding information overload (Vieira et al. 2016). Several adoptions of the BSC
towards sustainability addressing social and ecological dimensions have been
successfully implemented in the recentlast years (Figge et al. 2002; Arnold et al.
2003). However, the implementation of an energy viewpoint is hardly dealt with in
scientific literature (Vieira et al. 2016) or in practice (Laue 2016).

Based on the results of the case analysis an Energy BSC was modeled using the
software "ADOSCORE" from BOC. A typical corporate objective was selected to
demonstrate the usefulness of managing multi-dimensional corporate goals

including energy perspectives. "Increasing the number of park visitors (customers)" is a typical business goal and reflects the interdependencies between customer-, financial- and energy dimensions. The relations between these dimensions illustrates the impact of customer measures on financial and energy goals and vice versa. The energy BSC is illustrated in Fig. 3.

Pyramids with connected KPIs enable an ongoing controlling process and symbolize objectives in each perspective. Red or green dots signal a positive or negative development according to pre-quantified goals. An aspired increase of visitors (green dot) leads in the financial perspective to a revenue increase and higher profits which result in a rise of the equity ratio. Simultaneously on the energy perspective due to higher power demands through e.g. higher catering activities, the KPI for decentralized renewable power generation is turning negative as well as the

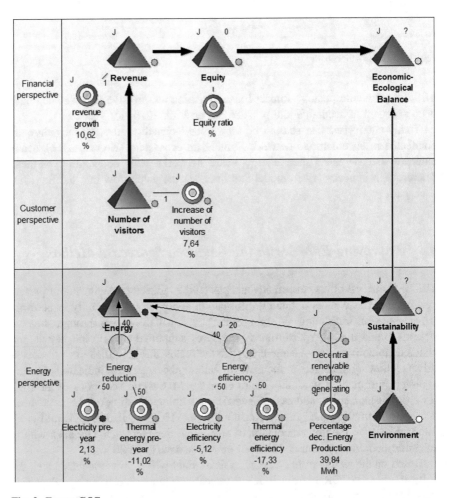

Fig. 3 Energy BSC

energy objective "Energy savings". The Energy BSC signals the energy manager, and respectively the board of management, when there is a need for adjustments.

The implementation of an Energy BSC showed that the method provides corporate management with a transparent controlling process integrating energy goals into corporate strategy. Our case model indicates direct negative consequences on the energy goals through increasing visitor numbers. Identifying such goal conflicts between strategic dimensions enables the management to set up compensating energy measures or to adjust customer goals.

For an effective EnMS an on-going controlling and decision making process with up-to-date information based on a data warehouse concept is necessary. However, the studied case expresses a gap regarding data tracking, based on heterogeneous data formats, and non-existing digital data flows that significantly hinder the multi-dimensional controlling process.

The BSC offers the possibility to manage the multi-dimensionality of a corporate energy system. The transfer to the energy domain is possible and results in a positive impact. Yet, the findings of the case study indicate clearly that a fully digital energy data process is the necessary starting point of effective and efficient BSC usage.

4.2 Modelling CDES with Enterprise Architecture Management

An enterprise is a complex and highly integrated system consisting of processes, organizations, information and technologies, with interrelationships and dependencies in order to reach common goals (Lankhorst 2013). A common problem of many medium-sized enterprises is the diverse IT-landscape. A mostly unsystematic growth of applications in enterprises over time results in "accidental architectures" (Giroti 2009). The corporate energy system, with its high diversity and its unsystematic growth, leads to a similar development with similar difficulties. Taking the case study results (criteria "Tracking"), data heterogeneity and non-existing energy data processes resemble the IT-domain. Therefore enterprises need to be aware of their relations among strategy, business processes, applications, information infrastructures and roles.

Enterprise Architecture Management (EAM) contributes to solve these problems. It is a holistic approach providing methods and tools to establish a complete perspective for enterprises (Lankhorst 2013). In this context, EAM provides an approach for a systematic development of organizations in accordance with its strategic goals (Ahlemann et al. 2012). Thus, EAM has evolved as a best-practice method that positively assists an EnMS. Additionally Appelrath et al. (2012) emphasized that changes of the energy system require an analysis of the corporate information systems as well.

In the findings of the case analysis, several characteristics were identified indicating EAM solution potential: (i) the complex, unsystematic development of corporate energy architecture; (ii) the different energy consumer units (e.g. catering facilities, greenhouses, event facilities), and (iii) the heterogeneous energy data landscape (missing data recording spots, analog data only, highly aggregated, useless data, various metering spots without an interconnected digital data workflow).

For modelling the Enterprise Architecture, the ArchiMate 3.0 (The Open Group 2016) modeling language was selected because the entity "physical elements" enables the modeling of power generation sites. The modeled Energy Enterprise Architecture (Fig. 3) shows a simplified representation of the case study energy system. The model enables the visualization of the as-is energy data process, the identification of digital gaps, and the planning of a roadmap towards a better fitting, future to-be Energy Enterprise Architecture.

The Energy EA (Fig. 4) reflects the business case "increasing park visitors" and its resolution in higher energy demand in catering facilities (Chap. 4.1). It describes the data sources for generated and consumed electrical power linked to side-by-side existing Excel-files. Today, the energy data recording process is carried out

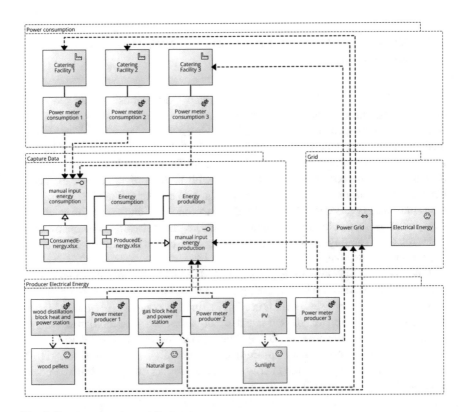

Fig. 4 Energy enterprise architecture

manually. The Energy EA displays the decentralized power generation system with its generation material as well as the catering electricity consumers. Producer and consumers are connected via the power grid. The Energy EA enables the energy manager to identify data gaps, existing data flows, the quality of data (analog-to-digital), data sources etc. This information is the baseline for planning the to-be-model based on a corresponding roadmap.

Modelling an Energy EA provides a positive impact for an effective CDES and is therefore a useful method for managing the energy domain in a digital, data driven process. Until now the modeling approach is simplified and represents only a small part of the CDES. Different viewpoints have to be integrated as well. Still, designing an Energy EA and the development of an EA-roadmap is the basis for the implementation of an effective Energy Scorecard.

4.3 Value Network Approach

Today, enterprises do not compete among individual companies, but among networks of interconnected enterprises (Peppard and Rylander 2006). Moore (1996) defines a business ecosystem as an "economic community supported by a foundation of interacting organizations and individuals". The performance of an enterprise is influenced, or even determined, by the performance and behavior of its partners. Therefore, it is necessary to analyze the cooperating partners and the exchanged values. The focus of a value network (VN) is to collectively determine value for each party involved (Al-Debei et al. 2013). This trend can be seen in the energy sector as well (Hertweck 2016; Küller et al. 2015).

Our case with its multiple-actors-landscape verifies this point too, so it is reasonable to examine and model the value exchanges between the different partners. A modeling approach including multiple actors is the e3v-method (Gordijn 2002). It enables the visual representation of value exchanges between VN actors by modeling their interactions. The diagram (Fig. 5) shows the actors of the ecosystem arranged around our case file (energy supplier, photovoltaic supplier, grid manager, wood pellets supplier). Several similar actors are aggregated into market segments (customer). Lines between the actors and market segments represent the value exchanges between the actors. These exchanges can be energy, money, energy generation sites, rent, governmental grant, material or data. Actors that do not have a direct value exchange are not pictured.

Our case study company achieves its CDES with the help of the PV- and the energy supplier. In return the company offers its building roofs to the PV suppliers. Each partner brings its competence into the VN which add up to the overall value. The modeled VN is also the starting point for describing the needed IT-services or data models between the actors and for expanding the Energy EA towards an Energy Network Architecture. Currently, digitized energy data hardly flow between the network actors. Paper-based billing procedures exist and an online-energy-tool to monitor the energy demand of the company is set up by the energy supplier. Yet,

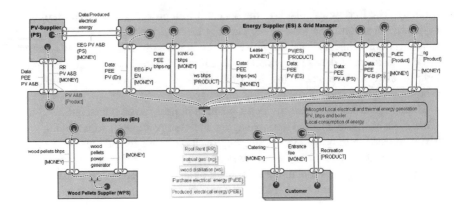

Fig. 5 Value network

the retrieved data hardly matches the energy information needs of the company. Therefore, a better alignment of energy information between the two actors is necessary. Additionally, it is currently not possible to automatically retrieve the energy information into the CDES. No digital energy information exchange exists at all between the PV supplier and the company.

Modeling the VN clearly has a positive impact on the company's transformation towards a digitized CDES regarding the role and the importance of each actor as well as providing a starting point for establishing an Energy Network Architecture.

5 Conclusion

The changing role of enterprises to and the establishing of CDES has risen constantly in recent years. Establishing an effective and efficient EnMS depends on a structured transformation path with multi-methods support. Digitization of energy data and processes is fundamental for a successful transformation. The success of managing a decentralized EnMS relies on nine categories. Applying these to our conducted case, three main challenges were identified: Managing multi-dimensional target systems, energy enterprise complexity and relying on an energy value network.

Three practice-proven methods that accompany the transformation path, the Balanced Scorecard, Enterprise Architecture Management and the Value Network approach were identified and evaluated in this paper. The BSC enables enterprises to manage complex multi-dimensional strategies. However, the current energy EA can't deliver the necessary energy data for proper system management. EA Modeling enables the visualization of the energy enterprise architecture to identify gaps in data flow and digital processes. These gaps define the roadmap towards a future digitalized Energy Enterprise Architecture that copes with the planned future development. The Value Network Approach displays dependencies with other actors and shows additional value opportunities. Additionally it indicates the interdependencies of the business models of each actor. All three methods clearly show the need for digitization to accomplish energy goals.

The transfer and partial integration of BSC, EAM and VN to the energy domain seem to offer a promising impact for transforming the corporate energy system. Further research establishing a methodology concerning the transformation path towards a CDES is required. Standardized interfaces between business demands and necessary energy system data sources are necessary. Conceptual models and their validation by empirical use cases will elaborate the data driven management of CDES.

References

Ahlemann F, Stettiner E, Messerschmid M, Legner C (2012) Strategic enterprise architecture management. Springer, Heidelberg

Al-Debei MM, Al-Lozi E, Fitzgerald G (2013) Engineering innovative mobile data services: developing a model for value network analysis and design. Bus Process Manag J 19:336–363. doi:10.1108/14637151311308349

Appelrath HJ, Terzidis O, Weinhardt C (2012) Internet of energy: ICT as a key technology for the energy system of the future. Bus Inf Syst Eng 4:1–2. doi:10.1007/s12599-011-0197-x

Arnold W, Freimann J, Kurz R (2003) Sustainable Balanced Scorecard (SBS): integration von Nachhaltigkeitsaspekten in das BSC-Konzept. Control und Manag 47:391–401. doi:10.1007/BF03254211

Baskerville RL (1999) Investigating Information Systems with action research. Commun Assoc Inf Syst 2:1–32. http://www.cis.gsu.edu/~rbaskerv/CAIS_2_19/CAIS_2_19.html

Christensen CM, Rosenbloom RS (1995) Explaining the attacker's advantage: technological paradigms, organizational dynamics, and the value network. Res Policy 24:233–257. doi:10.1016/0048-7333(93)00764-K

DIHK—Deutscher Industrie- und Handelskammertag, VEA—Bundesverband der Energie-Abnehmer e.V (2014) Faktenpapier Eigenerzeugung von Strom

Doleski OD (2016) Utility 4.0

Döring S (2015) Energieerzeugung nach Novellierung des EEG. Springer, Heidelberg

Figge F, Hahn T, Schaltegger S, Wagner M (2002) The sustainability balanced scorecard—linking sustainability management to business strategy. Bus Strateg Environ 11:269–284. doi:10.1002/bse.339

Flyvbjerg B (2006) Five misunderstandings about case-study research. Qual Inq 12:219–245. doi:10.1177/1077800405284363

Giroti T (2009) Integration roadmap for smart grid: from accidental architecture to smart grid architecture

Gordijn J (2002) Value-based requirements Engineering: Exploring innovatie e-commerce ideas. Vrije Universiteit Amsterdam

Hanschke I (2012) Enterprise-Architecture-Management—einfach und effektiv: ein praktischer Leitfaden für die Einführung des EAM. Hanser

Hauder M, Roth S, Schulz C, Matthes F (2014) Agile enterprise architecture management: an analysis on the application of agile principles

Hertweck D (2016) Methoden zur Entwicklung digitaler Serviceinnovationen. In: Lang M (ed) CIO-Handbuch. Strategien für die digitale Transformation. Symposion Publishing GmbH, Düsseldorf, pp 241–284

Hubbuch M (2016) Energy management in public organisations. In: Nielsen SB, Jensen PA (eds) Research papers for EuroFM's 15th research symposium at EFMC2016. Kgs. Polyteknisk Boghandel og Forlag, Lyngby, pp 172–184

Kaplan RS, Norton DP (1992) The balanced scorecard: measures that drive performance. Harv Bus Rev 70:71–79

Kaplan RS, Norton DP (1996) Using the balanced scorecard as a strategic management system. Harv Bus Rev 74:75–85. doi:10.1016/S0840-4704(10)60668-0

Koppenhöfer C, Fauser J, Hertweck D (2017) Digitization of decentralized corporate energy systems: supportive best-practiced methods for the energy domain. In: Rossmann A, Zimmermann A (eds) Lecture notes in informatics. Böblingen, Bonn, pp 91–106

Krcmar H (2011) Business model research—state of the art and research agenda. München

Küller P, Hertweck D, Krcmar H (2015) Energiegenossenschaften—Geschäftsmodelle und Wertschöpfungsnetzwerke. In: Zimmermann A, Rossmann A (eds) Digital enterprise computing 2015 (LNI). Gesellschaft für Informatik, Böblingen, Bonn, pp 15–26

Lankhorst M (2013) Enterprise architecture at work. Springer, Heidelberg

Laue H (2016) Die Umwelt geht vor. Solarpark, Ökogas & Co.—Ensinger Minaral-Heilquellen wirtschaften nachhaltig. Die Getränkeindustrie 2:16–18

Manthey C, Pietsch T (2013) IT-gestütztes Ressourcen- und Energiemanagement. Springer, Heidelberg

Moore JF (1996) The death of competition: leadership and strategy in the age of business ecosystems. HarperCollins, New York

Natural Resources Canada (2015) Energy management best practices guide. http://www.nrcan.gc. ca/sites/www.nrcan.gc.ca/files/oee/files/pdf/publications/commercial/best_practices_e.pdf. Accessed 3 Mar 2017

Peppard J, Rylander A (2006) From value chain to value network: insights for mobile operators. Eur Manag J 24:128–141. doi:10.1016/j.emj.2006.03.003

Posch W (2011) Ganzheitliches Energiemanagement für Industriebetriebe. Gabler, Wiesbaden

Razavi M, Aliee FS, Badie K (2011) An AHP-based approach toward enterprise architecture analysis based on enterprise architecture quality attributes. Knowl Inf Syst 28:449–472. doi:10. 1007/s10115-010-0312-1

The Open Group (2016) ArchiMate 3.0 Specification. 181

U.S. Department of Energy (2017) Federal sustainability/energy scorecard goals. https://www. energy.gov/eere/femp/federal-sustainabilityenergy-scorecard-goals

Uslar M, Specht M, Dänekas C et al (2013) Standardization in Smart Grids. Springer, Heidelberg

Van Grembergen W, Saull R (2001) Aligning business and information technology through the balanced scorecard at a major Canadian financial group: its status measured with an IT BSC maturity model. In: Proceedings of the 34th Annual Hawaii International Conference on System Sciences. IEEE Computer Socity, p 10

Vieira R, ODwyer B, Schneider R (2016) Aligning strategy and performance management systems: the case of the wind-farm industry. Organ Environ 1–24. doi:10.1177/ 1086026615623058

Solar Cadaster of Geneva: A Decision Support System for Sustainable Energy Management

G. Desthieux, C. Carneiro, A. Susini, N. Abdennadher, A. Boulmier,
A. Dubois, R. Camponovo, D. Beni, M. Bach, P. Leverington
and E. Morello

Abstract Cities play an increasingly important role with regards to energy transition. Main goal is to reach international and national (Swiss) targets related to energy efficiency and CO_2 emission reduction. As a contribution to these global challenges, during the last 6 years the State of Geneva has been producing a detailed solar cadaster. In order to facilitate periodical updates of this solar cadaster, the iCeBOUND project was launched. Around 10 public and private stakeholders, all linked within the Geneva Territorial Information System (SITG), collaborated on the project. Its aim was to design and develop a cloud-based Decision Support System (DSS) that leverages 3D digital urban data with high computing performance, hence facilitating environmental analyses in large built areas, like solar energy potential assessment. As result of the project, an official geoportal and a newfangled public web interface were made widely available early 2017, so as to strengthen decision making with regards to solar installation investment.

Keywords Energy management · Solar cadaster · Urban digital models · Cloud computing · Official geoportal · Public web interface

G. Desthieux (✉) · C. Carneiro · N. Abdennadher · A. Boulmier · A. Dubois · R. Camponovo
University of Applied Sciences Western Switzerland (HES-SO/HEPIA), Geneva, Switzerland
e-mail: gilles.desthieux@hesge.ch

C. Carneiro
e-mail: cmagcarneiro@gmail.com

A. Susini · P. Leverington
Geneva Energy Agency (OCEN), Geneva, Switzerland
e-mail: alberto.susini@etat.ge.ch

D. Beni · M. Bach
Arx-IT, Geneva, Switzerland

E. Morello
Politecnico di Milano (POLIMI), Milan, Italy

© Springer International Publishing AG 2018
B. Otjacques et al. (eds.), *From Science to Society*, Progress in IS,
DOI 10.1007/978-3-319-65687-8_12

1 Introduction and Motivation

During the last years, urban studies emphasized that cities are playing a leading role with regards to environmental issues and energy transition. In 2014, EU countries have agreed on a new 2030 Framework for climate and energy, including EU-wide targets and policy objectives for the period between 2020 and 2030. One of the targets involves reaching at least a 27% share of renewable energy consumption (European Commission 2014).

In order to contribute to such global targets, cities need to improve their capability to evaluate local renewable energy sources available, thus promoting and generalizing their use. Related to this topic, solar energy is a major driver of energy transition. Therefore, it is highly relevant to target house and building roofs located in urban areas for producing solar energy (both thermal and photovoltaic). From 2011 to 2016 a step-by-step solar cadaster was developed for the State (Canton) of Geneva. It was based on the use of LiDAR data for 3D urban models construction, solar modelling, computational models and digital image processing techniques. First stage (2011) permitted to make publicly available a new solar cadaster emphasizing raw solar radiation on building roofs. Second stage (2014) allowed to extract several rooftop indicators related to energy production, economical investment/payback and environmental assessment. Third stage (2016) privileged an update of 2011 solar cadaster using new LiDAR data, 3D urban models, refined solar modelling algorithms and improved computational models. Updated outputs are displayed on both the Geneva official geoportal (SITG—Geneva Territorial Information System) and a novel public web interface (please consult Sect. 4). The geoportal,[1] mainly addressed to professionals of solar energy, has the ability to extract the whole solar database (as open data) on any perimeter or group of buildings, hence allowing to support solar energy planning at different scales. This is often not the case for other geoportals, such as the Swiss solar register,[2] to be accomplished in 2018. Its goal is to provide nationwide data about solar energy potential, but only for each building roof at once.

Moreover, most of urban solar cadasters widely available in Europe are made on rather restricted areas, generally less than a hundred square kilometers. Differing from this trend, Geneva's solar cadaster presented in this paper is not limited to the city of Geneva but instead covers the whole State of Geneva, reaching an area of around 300 km². Such an area is challenging with regards to computation time. For instance, first Geneva's solar cadaster (2011) calculation, computed by a single server machine hosted at the University of Applied Sciences, Western Switzerland (HES-SO/HEPIA), Geneva, took around 2000 h.

Therefore, in order to boost computation time for forthcoming solar analysis on large built areas, cloud computing use has been identified as highly priority by all stakeholders involved in this project. For this purpose, iCeBOUND project,

---

[1]https://www.etat.ge.ch/geoportail/pro/?mapresources=ENERGIE_SOLAIRE.

[2]http://www.toitsolaire.ch.

supported by the Swiss Federal Agency for the Promotion of Innovation Based on Science (CTI[3]), was launched in 2013. Its goal was to design and develop a cloud-based Decision Support System (DSS) that puts forward the use of 3D digital urban data and state of the art computing skills in order to facilitate environmental analyses in cities, such as an assessment of solar energy potential.

In the framework of 2016 Geneva's solar cadaster updating and improvement, iCeBOUND project aimed at providing decision makers with relevant indicators for city planning and energy management. Two key goals of this project were to: (i)—improve the visibility of how much energy can be produced, by both photovoltaic and thermal technologies, thus allowing citizens to check the potential of their building; (ii)—promote the installation of solar panels and linked infrastructures, therefore giving the opportunity to increment business strengthened by solar energy and to boost energy transition from nuclear to "green technologies".

Firstly, this paper introduces methodological background used for solar resource modeling. Secondly, it describes cloud computing used for the computation of rooftop solar potential. Finally, it highlights end results achieved through the public web interface development.

2 Methodological Background: Solar Resource Modelling

The methodological background is based on previous works carried out by some of the authors of the present paper (Carneiro et al. 2009; Desthieux et al. 2011, 2014). "Solar Energy Potential" consists of evaluating the potential of building roofs located in urban areas for producing solar energy using both thermal and photovoltaic technology. Hence, the creation of solar maps of cities allowing to accurately evaluate areas that can be used for the installation of renewable energy, such as solar panels on building roofs, is currently considered as highly relevant.

Solar energy potential is calculated by summing the estimated value of the solar radiation in kWh for every hour during a time span at a given place. It is computed using: location (latitude); average solar radiation parameters; 3D surrounding relief and landscape; hourly shading (at a given position); ratio of the visible sky (at a given position), also called "Sky View Factor" (SVF). The three first points above are input data, whilst the two last are calculated using a Shading Algorithm. Data needed for processing is stored in georeferenced TIFF raster format (GeoTIFF), presented as follows:

1. Digital Surface Model (DSM), here entitled 2.5-D Digital Urban Surface Model (2.5-D DUSM), describes the elevation on each point (pixel) of a raster image

---

[3]Among other things, CTI aims to bring cutting-edge companies closer to academic research, thus allowing to create a strong dynamic in R&D promotion and to enhance the potential for innovation. Involved partners are Geneva energy agency, Geneva facility agency-SIG, Geneva geomatics agency, Geneva topographic, agency, CTI, Arx iT, HEPIA, CERN, EPFL and POLIMI.

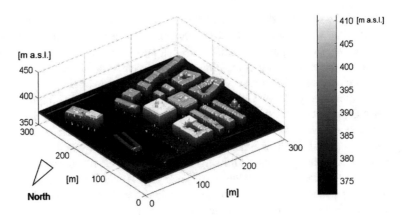

Fig. 1 Example of a digital surface model (2.5-D DUSM) for a neighborhood of Geneva: meters above sea level (m.a.s.l.) are illustrated in a *grey scale*

with specific resolution (Fig. 1). It is constructed from LiDAR data and represents the structure of the city, such as buildings and houses.

2. The slope matrix, produced from 2.5-D DUSM using common GIS software tool, describes the slope of each point (value between 0 and 90°).
3. The orientation matrix, produced from 2.5-D DUSM using common GIS software tool, describes the orientation of each point (value between 0 and 360°).

Solar radiation is calculated by summing its three main components: direct radiation, direct use radiation and reflected radiation. Direct radiation is directly proportional to sun visibility; reflected radiation depends on the ground and the nearby object reflection; diffuse radiation is derived from sky visibility. The used shadow casting routine is based on the algorithms developed by Ratti and Richens (2004). It calculates the shadow map of a given 2.5-D DUSM according to a given light source. Therefore, this map contains boolean values that represent the shading of each point (0 for shaded, 1 otherwise) for a given light/signal source position.

According to Fig. 2, the purpose of Shading Algorithm is to assess, for each point (pixel) P_0 that belongs to a 2.5-D DUSM, which other point P_1 is shading it according to a given light/signal source. Figure 2 shows the algorithm explained above. When the light source is in position 1, H_1 is higher than P_1 and P_0 is lighted. On the other hand, when the light source is in position 2, the height H_2 is less than that building (point P_1) and P_0 is shaded. The Shading Algorithm uses three input data: light/signal source position (azimuth and altitude); 2.5-D DUSM and latitude of the location of the given model. The output of this algorithm is a shadow map that contains the shading state of each point that belongs to the 2.5-D DUSM under analysis.

The value of direct solar radiation at P depends on shading algorithm for a given sun position (azimuth and altitude). Thus, if P is shaded its value is zero, one otherwise. Figure 3 shows three calculated shadow maps for an area of Geneva at 4 pm, 5 pm and 6 pm (September 15th).

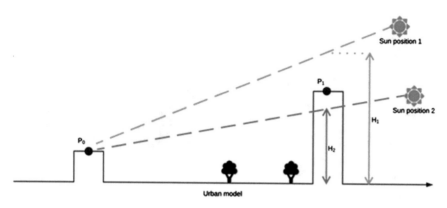

Fig. 2 Principle of the shading algorithm

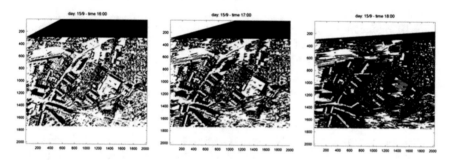

Fig. 3 Calculated shadows for a neighborhood of Geneva at three times of the same day

The sky visibility (or "Sky View Factor: SVF") is used to compute shadowing on diffuse radiation. SVF describes the amount of visible sky from a point P—it is calculated by using a sky model. The latter represents the sky as a vault composed of different light sources. As for the sun, a light source position is given by its azimuth and altitude component. For instance, a point located in an urban area can have SVF near zero; whereas a point located in a rural area mostly have SVF close to 1.

For the SVF calculation, a sky vault with 580 light sources is used as input data, based on the Tregenza sky subdivision model (Tregenza 1987). It involves 580 shadow processes as Shading Algorithm is repeated the same number of times.

To sum up this methodological section, solar radiation calculation is based on four distinct processes and data (Fig. 4):

1. Raw data collection (heights information by remote sensing and building or rooftop footprints);
2. Raster mask inputs for solar modelling (pre-processing): building of 2.5-D DUSM from height data and its derived masks—roof cell slope and orientation. Subdivision of the large studied areas in tiles as processing unit (please consult Sect. 3);

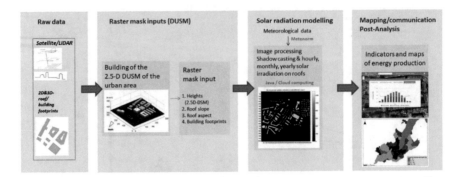

Fig. 4 Whole process of solar modelling and mapping from raw data acquisition to solar mapping

3. Use of "Meteonorm" historical worldwide database for assessing solar radiation along the urban fabric. It provides radiation values (global, diffuse and normal) with a time step of 1 h;
4. Solar radiation modelling: (a) SVF calculation based on 2.5-D DUSM and the sky subdivision model of 580 light sources, (b) hourly shading calculation based on the 2.5-D DUSM and sun positions as input data (one shadow process per hour during for establishing the direct sun visibility), (c) for every cell of the 2.5-D DUSM, computation of hourly solar radiation, based on statistical meteorological data of the studied area (hourly global, diffuse and direct solar radiation on flat surface), integrating shadow (SVF and direct) from outputs a) and b), as well as slope and aspect of the cell (based on 2.5-D DUSM). As a result, monthly and yearly solar radiation values are calculated for each pixel of the outputted raster grids. An example of yearly and monthly irradiation values (kWh/m^2/year) calculated for Geneva's downtown is illustrated in Fig. 5. Pixel-based solar radiation values are then aggregated by vector polygons (e.g., roofs and parts of roofs);
5. Mapping and communication (post-processing): from outputs of solar modelling, calculation and mapping of indicators on building roofs, and communication through the official geoportal and public web interface (please consult Sect. 4).

The outputs from the method proposed here were scientifically validated by an international expert on solar modelling (Ineichen 2012).

3 Cloud Computing (iCeBOUND Project)

Shading Algorithm is observed to be an irregular and time consuming algorithm. Moreover, SVF calculation executes this algorithm several times to provide accurate information. Hence, this algorithm was parallelized in the cloud infrastructure in order to be more efficient and less time-consuming.

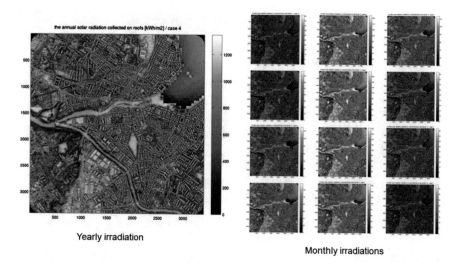

Yearly irradiation

Monthly irradiations

Fig. 5 Yearly (*left*-hand side image) and monthly (*right*-hand side image) irradiation values (KWh/m2) calculated for Geneva's downtown

The system is based on a CPU and memory intensive "shadow process" algorithm. During iCeBOUND project two paradigms were used to describe this algorithm. The first paradigm is based on a conventional parallelization. The second one is based on an optimized data distribution. The two paradigms are implemented on cloud computing infrastructures. Computational design and architecture employed use the Cloud infrastructure hosted in HEPIA. Nevertheless, other cloud systems like Amazon could be also used. For more information on technological specifications of the Cloud infrastructure, the reader can refer to Boulmier et al. (2016a, b).

Despite the use of cloud computing significantly accelerates calculation time, as total area of the State of Geneva is too wide, it is not conceivable to make all calculation at once. Therefore, the 2.5-D DUSM of the State of Geneva was divided in about 55 tiles of 3.4 × 3.4 km, including an additional buffer of 200 meters from each side in order to take into account mutual shadowing from one tile to another (Desthieux et al. 2011). A regular pre-processing tiling grid of 2.5-D DUSM is applied to each studied area in order to compute solar radiation tile by tile.

4 End Results: Public Web Interface

A specific public web interface[4] (Fig. 6) is provided in order to communicate about the solar energy potential of the State of Geneva's building roofs. In particular, maps and information are published in an interactive way for different types

---

[4]https://sitg-lab.ch/solaire/.

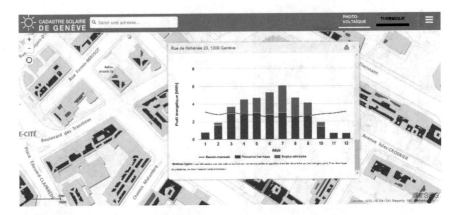

Fig. 6 Geneva's public web interface: example of display for main rooftop characteristics

of end-users (e.g. authorities for energy planning, energy companies for marketing and investing, and public for awareness and encouragement in solar PV installing).

Main functionalities expected for the public web interface are presented as follows:

- Identification of suitable roof parts for thermal and PV panel installation considering yearly radiation thresholds (e.g. 1000 kWh/m^2/year);
- Detailed assessment of rooftop solar energy potential for thermal and PV technology, in terms of available suitable rooftop area (m^2), total installed capacity (MW), total potential generation (MWh) and economic and environmental indicators (saved CO_2 emission);
- Visualization of maps with color classification related to electric solar map potential and display of graphics with regards to energy and money savings by rooftop;
- Make viewable (afterward user request) information about main rooftop characteristics, such as orientation, slope, available area, etc., for initial design of thermal and PV installation projects;
- Generation of summary PDF and export data to CSV format (or XML).

5 Conclusion and Perspectives

A decision support system (DSS) based on cloud computing has been developed with iCeBOUND project, thus allowing to accurately assess the effective thermal and PV technology production for different end-users, such as citizens and solar panels installers, and support decision making on launching installation projects. The system quantifies the solar power that a roof can generate by taking into account local weather parameters and 3D urban digital data, thus allowing to evaluate how much shade falls on each roof from nearby trees and buildings.

Moreover, Geneva's solar cadaster is made publicly available through both the official geoportal (extraction of open data for professionals) and the public web interface (general public awareness). Sharing data and information this way is indeed crucial for promoting an open Swiss energy transition strategy from nuclear to "green technologies". Hence, available data strengthen important decision with regards to solar installation investment: they help design projects and provide to real estate owners financial evaluation before applying for public incentives and contacting companies specialized on solar panel installation. The monitoring of the public web interface shows between 80 and 100 consultations per week, which is mostly in line with the number of applications for public funds (in particular in the framework of the Geneva funding program for renovation of buildings).

Next step will be to: (i) develop a wide-ranging communication to potential stakeholders: remarks and ideas should be collected for enhancing the public web interface in line with their needs; (ii) extend the solar cadaster to the whole agglomeration of Geneva, which includes French municipalities (a cross-border solar cadaster is undoubtedly a challenge in terms of energy management); (iii) speed up calculation time through the use of novel GPU (*Graphics Processing Unit*) accelerated computing.

Acknowledgements Authors are definitely indebted to several Geneva institutions that truly contributed to iCeBOUND project: L. Niggeler and G. Cornette (Geneva topographic agency), P. Oehrli and A. Vieira de Mello (Geneva geomatic agency), C. Anthoine-Bourgeois, J. Shulepov and J. R. Fahrni (Geneva facility agency, SIG).

References

Boulmier A, Abdennadher N, Desthieux G, Carneiro C (2016a) Cloud based decision support system for urban numeric data. Final report. http://lsds.hesge.ch/wp-content/uploads/2016/09/iCeBOUND_Final_Report.pdf

Boulmier A, White J, Abdennadher N (2016b) Toward a cloud based decision support system for solar map generation. In: IEEE 2016—8th international conference on cloud computing technology and science, pp 230–236

Carneiro C, Morello E, Desthieux G (2009) Assessment of solar irradiance on the urban fabric for the production of renewable energy using lidar data and image processing techniques. In: Sester M, Bernard L, Paelke V (eds) AGILE 2009—12th conference, hannover. Advances in GIS. Springer, Heidelberg, pp 83–112

Desthieux G, Carneiro C, Morello E (2011) Cadastre solaire du Canton de Genève. Rapport final. Study done for the Geneva Canton

Desthieux G, Gallinelli P, Camponovo R (2014) Cadastre solaire du Canton de Genève,—Phase 2; Analyse du potentiel de production électrique par les panneaux solaires thermiques et photovoltaïques. Rapport final, Study done for the Geneva Canton

European Commission (2014) Communication from the commission to the European parliament, the Council, the European economic and social committee and the committee of the regions, A policy framework for climate and energy in the period from 2020 to 2030 [COM(2014) 15]

Ineichen P (2012) Projet potentiel solaire: validation des modèles. Report. https://archive-ouverte.unige.ch/unige:23641

Ratti C, Richens P (2004) Raster analysis of urban form. Environ Plan 31:297–309

Tregenza PR (1987) Subdivision of the sky hemisphere for luminance measurements. Lighting Res Technol 19:13–14

Goal-Based Automation of Peer-to-Peer Electricity Trading

Jordan Murkin, Ruzanna Chitchyan and David Ferguson

Abstract As the uptake of microgeneration increases, the centralised model of electricity generation will be significantly altered. One of the new models that is being investigated is the notion of a peer-to-peer electricity market in which prosumers can market their electricity exports to any other household. This paper investigates this model and proposes a new algorithm for automating the sale and purchase of electricity in this market aiming to optimise the market while providing increased control to householders.

Keywords P2P electricity trading · Trading automation · Goal optimisation

1   Introduction

Microgeneration is the generation of small amounts of electricity by individual households (e.g., via solar panels) and non-energy specialist small businesses (e.g., farmers via wind turbines). Uptake of the microgeneration in the UK is steadily increasing (DEC 2016) and with it new market models are necessary to make full use of this newly available generation capability. One of the new models that has gained a lot of research and practical interest with increased *distributed generation* (i.e., generation by other than centralised power plants) is the concept of *peer-to-peer (P2P) electricity trading* (Power Ledger 2016; Power Matcher 2017; Brooklyn

J. Murkin (✉) · R. Chitchyan
Department of Informatics, University of Leicester, Leicester, UK
e-mail: jpm45@le.ac.uk

R. Chitchyan
e-mail: rc256@le.ac.uk

J. Murkin · D. Ferguson
EDF Energy, London, UK
e-mail: david.ferguson@edfenergy.com

© Springer International Publishing AG 2018
B. Otjacques et al. (eds.), *From Science to Society*, Progress in IS,
DOI 10.1007/978-3-319-65687-8_13

Microgrid 2017) in which excess microgeneration can be sold by the producer directly to another consumer instead of to the grid.

A P2P market model would bring a number of benefits over the current system; prosumers would be able to market their excess generation, possibly providing an extra income stream. While consumers would see a far greater choice over their energy source, with the ability to pick the exact source of their supply. However, should the participants be required to manually manage sales, the required time commitments and need to acquire trade-related domain knowledge would likely outweigh any benefit for the majority of users. Consequently, it is necessary to automate the trading process, and with automation also comes potential optimisation.

This paper will explore the methods of automation with the objective to provide increased choice to consumers and incentivise the purchase of locally generated electricity to increase the value proposition of microgeneration. The intent being to reduce transmission losses caused by long distance energy distribution and increase the use of renewable technologies, both aiming to increase energy efficiency in the energy industry. Thus the contribution this paper provides is to propose:

- An electricity market structure that enables the automation of P2P trading;
- A new algorithm to automate P2P electricity trading;
- Evaluation of performance of this algorithm using simulations.

This paper investigates the different ways in which electricity trading over a P2P market could be automated and the different optimisations it makes possible. It will start with a background of the current research that has taken place in this area, the ways that a peer-to-peer market may appear and ends by proposing a new matching algorithm that could be used to optimise this future market.

2 Background

Previous research has explored different methods of household trading inside the electricity market. The authors of (Yaagoubi and Mouftah 2015a, b) used a modified regret matching as a method to match buyers and sellers individually, allowing for price preferences and taking into account transmission losses in the network. However, in (Yaagoubi and Mouftah 2015a) each buyer can only receive energy from one seller and price is the only consideration. Another game theoretic approach was used in (Wang et al. 2014) which focused primarily on the increased utilisation of energy storage using an auction based approach to determine price. A microgrid focused trading mechanism (Luo et al. 2014) has also been suggested that relies on sharing energy schedules between participants so as to match producers and consumers at the correct time to ensure electricity is utilised fully. This research expands on these with the addition of multiple preferences for each participant in the network, the ability for buyers to purchase from multiple sellers and vice versa and proposes a new market matching approach designed to find an optimal solution while reducing overheads.

While automation is the focus, it is important to first detail how the market would be structured and function in a P2P trading environment. The structure of the market will also inform the choice of trading algorithm type.

The current electricity market divides each day into 30 min periods which electricity can be traded for. This research kept the same approach in which trading is separated into clearly defined periods allowing for regular trading to continue as well as providing a structure to the automation. It is currently undecided whether these fixed trading periods will remain at 30 min or whether a shorter time period will be used. To decide this, a number of simulations will be run with different trading periods to identify if there is a benefit to shortening these periods.

2.1 Structure

To begin, it is necessary to define the structure of these trading periods and the trading procedure. In the current electricity market, trade contracts are determined in advance between parties and controlled by those involved. To remove the complexity of the electricity market and make it accessible to the average consumer, contracts will instead be organised by the market itself. This means that from a consumer's point of view the only interaction with the system is the setup of their account and the configuration of their preferences.

With the market establishing contracts between participants, there are two main approaches that can be taken. The first and more traditional is to automate the trades of each participant individually. The second is to have a single entity that performs a matching of all buyers and sellers together, negotiating all contracts at the same time. The first serves to benefit individuals to the maximum potential, but may not find the optimal solution for the network as a whole. The second allows for matching to be decided on a market level, reducing overheads. This method though, is generally more open to abuse by having a single entity controlling all contracts being created. As this research is built upon distributed ledger technology (DLT), as described in (Murkin et al. 2016), which supports trust by design, this risk of abuse can be avoided. For this reason, the second option was chosen. This left a number of choices available, summarised in Table 1.

The choice of which option to use would also depend on which algorithm would be used to match buyers and sellers together. There are many types of matching algorithm. In these (in general), sellers would announce what they have to sell and buyers would announce what they want to purchase. A system would then match buyers and sellers based on price and/or preferences, using such mechanisms as regret matching, sharing electricity schedules, auctions or rank order lists.

While option 4 supports a truly open market, where a product can be listed for any price and any buyer can instantly purchase that product, it does not apply well to electricity mainly due to network latencies playing such a large role in sales.

Table 1 Market design options

Option	Seller	Buyer	Market	Notes
1.	List sales	List buy orders	Match buy orders to sales	Overheads for matching all buyers against all sellers
2.	List sales	Directly assign buy orders to sales	Determine which buyer gets which sale	Reduced overheads for matching buyers against specific sellers
3.	List sales	Bid on sales	Bids are frozen at the end of trading and the highest bids win	High bidding overheads under large-scale trading
4.	List sales	Purchase sales	First come first served basis	Network latencies will have large effect on sales

An auction mechanism as described in option 3 would likely allow significant control to users of the system, but was removed due to the overhead that would be created if used on a large scale.

The final options, 1 & 2, are quite similar, but again due to possible increased overhead for minimal improvement, option 1 was removed and the project settled on a variation of rank order list matching to fit option 2.

This type of matching typically sees two groups, each member of each group has a list defining their preference over the parties in the other group. These two lists are then used to determine how the participants from each group are matched. This will be detailed further in the algorithm section.

With the trading process defined, it is then possible to structure the trading periods by splitting them into defined sections. These can be seen in (Fig. 1).

The purpose of these sections is described below:

Sell: In this period participants will place their sell orders.

Buy: In this period buyers would list interest against the sales created in the Sell period.

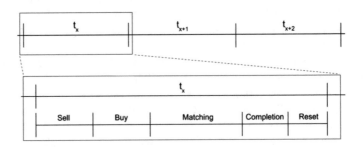

Fig. 1 Proposed structure for P2P trading periods

Matching: This period will deal with the matching aspect of the system, wherein buyers would be matched to specific sales.

Completion: This section has been added due to the DLT nature of this platform. Time is given to ensure that transactions are completed successfully.

Reset: This section again has been added due to the DLT nature of the platform. This section will reset all necessary fields and arrays ready for the next time period.

3 Algorithm Design

With a market structure in place and an algorithm picked for the matching procedure the next stage was to implement it. This section details the design decisions that took place while implementing this algorithm and the considerations that need to be made when automating trades for a P2P electricity market.

3.1 Considerations

An integral element of this system is the distance charge. It was added to the system to serve two purposes. Firstly, to pay for the infrastructure necessary for the electricity network to function including any other obligations that need to come through the sale of electricity. Secondly, to differentiate each sale from one another. Whereas in a traditional commodity market, from a consumer's point of view each unit is thought of to be identical to enable the market to function correctly. The distance charge serves the opposite purpose, intending to differentiate each sale. The reason for this is that while electricity is identical from all sources, transmission loss and other associated costs are largely dependant on the distance it travels.

3.2 Preferences

A primary objective of this research is to provide a mechanism to increase choice over electricity source. The current system already allows choice over price, but there is limited choice that accurately reflects real-time wholesale prices. Generation type is already partially supported with suppliers selling products that are backed by renewable sources. Distance is not supported at all, as in the current system electricity is seen as a commodity in which each unit is identical from a consumer's point of view.

To incorporate these into the algorithm, each participant is given a method to specify a number of preferences that allowed them to control the way in which they were matched by the market. The preferences included for sellers are (i) Minimum price preference and (ii) Distance preference. Preferences for buyers are

(i) Maximum price preference, (ii) Distance preference, and (iii) Energy type preference.

These preferences are used as a means to create scores between the market participants creating the rank order lists for each.

3.3 Scoring

The scoring module of the algorithm dictates how matches are made and so was designed as an independent module allowing it to be replaced whenever needed allowing easy testing of different calculation methods. Below, the calculations used for the current implementation are shown.

In this section a number of letters will be used throughout to improve readability of equations. Firstly, B and S will be used to represent buyer and seller respectively and these letters may be appended to other letters, such as Pb referring to buyer's price preference and Ps referring to sellers price preference.

Other letters:

P Price preference

D Distance preference

E Energy generation type

C_d Distance charge

EP() a function returning the energy type preference given an energy type E

d distance

The scoring system used needs to incorporate the preferences that were previously mentioned. For each of these preferences a different method was used, this section will go through each. To begin it is first necessary to calculate the distance between the participants to allow for the score to be calculated. This was implemented by adding a latitude and longitude to each participant and then using a great-circle calculation to determine the distance between the two points.

3.3.1 Price Preference

The price preference is used as a method to create an initial score between the participants. By using the price preference of both the buyer and seller and including the distance charge we can create a basic score that also reflects the maximum price that the buyer is willing to pay for the given sale.

$$score = P_b - (d \times C_d)$$

This current score only takes into account buyer preference. To incorporate seller preference as well, a check is added to ensure that the maximum price that a buyer is willing to pay is above the minimum price they are looking for. This means discarding all scores in which $score < P_s$.

3.3.2 Energy Type Preference

Allowing householders to specify the energy type they prefer provides a significant benefit over the current system in which electricity sources are indistinguishable. This preference allows individual preference to be given for each microgeneration energy type: solar, micro CHP, wind, hydro, and anaerobic digestion. Each of these individual preferences is normalised to 1, thus the energy type preference is included by multiplying the score as follows $score = score \times EP(E_s)$.

3.3.3 Distance Preference

To incorporate the buyer's distance preference a simple check is added to determine if the seller is outside of the buyer's specified distance. If the seller is farther than this distance, a multiplier is applied to the score so as to reduce the score more as the distance increases. The reason for this is to avoid creating an exact range within which a buyer searches for sales, instead allowing sellers outside to still be considered.

$$score = score, d < = D_b$$

$$score = score \times (D_b \div d), d > D_b$$

The same mechanism can then be applied to incorporate the seller's distance preference.

$$score = score, d < = D_s$$

$$score = score \times (D_s \div d), d > D_s$$

3.3.4 Final Score Calculation

After each section described above had been incorporated this gave a final method for calculating the score between two participants as:

$$score = (P_b - (d \times C_d)) \times EP(E_s), d < = D_b, d < = D_s$$
$$score = (P_b - (d \times C_d)) \times ((D_s \div d) + EP(E_s)) \div 2, d < = D_b, d > D_s$$
$$score = (P_b - (d \times C_d)) \times ((D_b \div d) + EP(E_s)) \div 2, d > D_b, d < = D_s$$
$$score = (P_b - (d \times C_d)) \times ((D_b \div d) + (D_s \div d) + EP(E_s)) \div 3, d > D_b, d > D_s$$

3.3.5 Implementation

As noted, our algorithm is adapted from a traditional rank order list matching. Here:

1. Seller exports are listed as sales
2. Buyers are assigned to all sales for which their preferences suit and a score is assigned between the two. This score acts as a method to create rank order lists for both the buyer and the seller. For buyers, this is the list of sales they are assigned with the respective scores for these sales. For sales this is the list of interested buyers and their respective scores.
3. The highest scoring buyer for each sale is assigned as an initial match to this sale.
4. The highest score of all initial matches for each buyer is chosen to realise a sale.
5. The algorithm then repeats from step 3 until there are no buyers assigned to a sale, no more sales or there is no remaining consumption to satisfy from buyers.
6. This algorithm performs an individual based optimisation but can easily be modified to incorporate other types of optimisation (e.g. a score maximising algorithm to allow for market level optimisation).

4 Preliminary Evaluation

To assess the performance of this algorithm a JavaScript simulation was implemented. As we aim to assess the performance of the algorithm (and not to accurately simulate an electricity market) a number of generators were created to automate the creation of the accounts and simulate their generation and consumption. Simulation configuration: The accounts were randomly generated for each simulation to ensure that the algorithm functioned correctly regardless of the accounts it used. They were assigned a random location inside an arbitrary area in the UK, with latitude between 50.956,870 and 52.438,562 and longitude between −2.386,779 and 0.292,914. A percentage of these accounts were assigned to be prosumers. Price preferences were randomised; for prosumers, minimum sell price set between 46p/kWh and for consumers maximum purchase price set between 14–16 p/kWh. Distance preference was set for all accounts between 5–10 km. Energy type preferences for consumers were randomised between 0.5–1 and then normalised to 1 between all preferences with the highest preference being set to 1. Lastly generation type for prosumers was randomised to reflect figures from UK microgeneration installations (DEC: Monthly Central Feed-in Tariff register statistics—Statistical data sets—GOV.UK 2016).

Simulations were run for 16 different scenarios, where the percentage of prosumers in the market ranged from 5 to 20% in 5% increments, and the number of participants ranged from 500 to 2000 in increments of 500. In all scenarios the prosumers generated between 5–10 kWh of electricity and consumers consumed between 1–6 kWh and the distance charge was set to 0.2 p/km/kWh. Each scenario was run 50 times and then averaged and the results can be seen below.

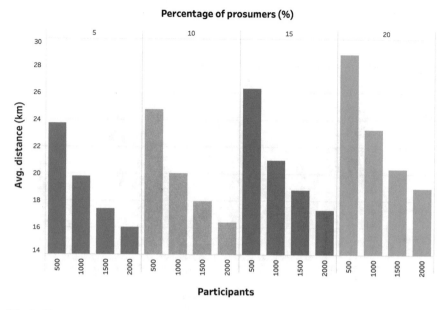

Fig. 2 The average distance between sales for each scenario

Simulation results: Here we assess the algorithm and its performance.

Figure 2 shows how the distance between sales changes through the different scenarios. It can be seen that an increase in the number of participants in the market decreases the average distance between sales. This is likely due to an increase in the density of participants and shows that the algorithm is performing as expected, incentivising the use of local generators.

A point to be noted is that as the percentage of generators increased, the distance between sales also increased. This was a more surprising result, but could be explained by a decrease in the density of buyers causing an increase in the average distance between buyers and sellers.

Another important characteristic of the algorithm can be seen in Fig. 3, which shows how the price per kWh is affected through each scenario. This figure shows that the algorithm reflects the standard economic model of supply and demand. As the number of participants (demand) increases, competition increases and this increases the average price paid. At the same time it can be seen that with an increased number of prosumers (supply) a decrease in the average price occurs.

Figures 4, 5 and 6 assess the performance of the algorithm. It can be seen that as the number of participants increases, the number of sales, the number of rounds and the time taken all increase linearly. The time taken though increases with a significant gradient and thus could pose a problem with a large number of participants. As the current simulation is running synchronously it may be possible to reduce through parallelisation and caching mechanisms to reduce calculations per round if it was to be scaled to millions.

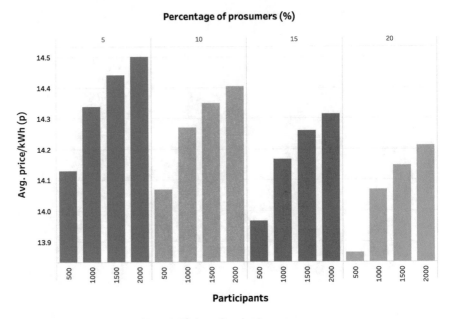

Fig. 3 The average price paid per kWh in each scenario

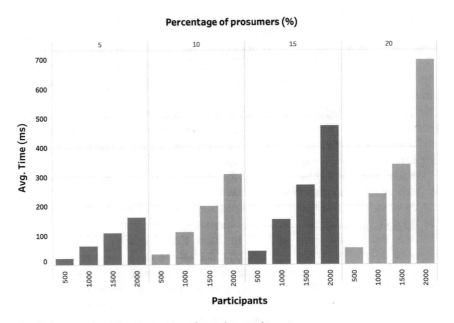

Fig. 4 Average time taken to converge for each scenario

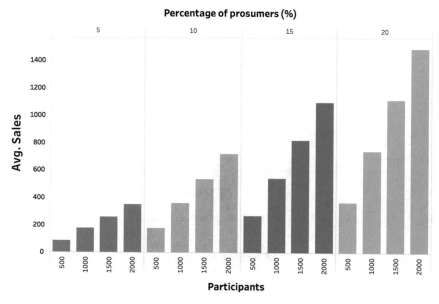

Fig. 5 The number of sales that took place in each scenario

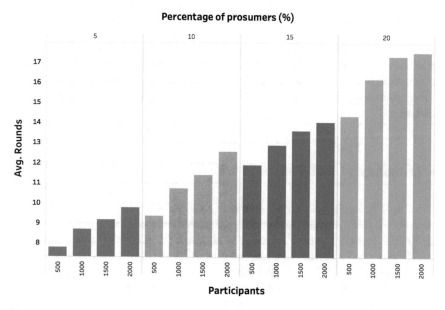

Fig. 6 Average number of rounds taken to converge for each scenario

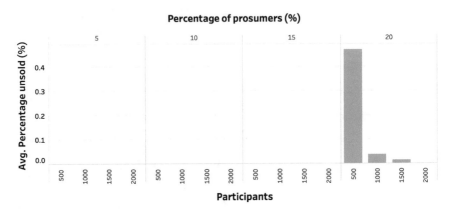

Fig. 7 Average percentage of unsold generation for each scenario

The final important piece of information is the amount of unsold generation throughout the simulations. With 5%, 10% and 15% of participants as prosumers there was no unsold generation, but as the share of prosumers increased to 20% a small amount of generation was unsold Fig. 7. Yet, as the number of participants increased further, this unsold generation again dropped to zero. This is likely due to the large area used in simulation, leading to large distance between participants, with some generation unsold due to high distance charge effect.

These evaluations allow us to conclude that the algorithm is functioning as originally intended, incentivising the purchase of locally produced electricity, minimising the electricity that is left unsold and encouraging competition. We have also seen that while at a low number of participants (thousands) the algorithm will converge in acceptable times, large scale use (millions) would likely require some alterations.

5   Conclusion and Future Work

This paper has presented a new approach to buyer and seller matching in a P2P electricity market. Through simulations we have shown that the algorithm performs acceptably, reducing unsold electricity to almost zero, incentivising the purchase of locally produced electricity and accounting for individual preferences, but may require alterations for large-scale implementation.

As this research continues: real locational data will be added to the simulation to help locate areas that would benefit most from this type of trading mechanism and remove the unrealistic spacing between households; and energy data will be added to conclude the quantitative benefits to both buyers and sellers that can be achieved using this algorithm.

The next stage of this research will investigate alternative scoring mechanisms and algorithms to look beyond purely individual optimisation to allow for group based optimisation and optimisation of the market itself.

References

Brooklyn Microgrid. http://brooklynmicrogrid.com/ Accessed 28 Apr 2017
DECC: Monthly Central Feed-in Tariff register statistics—Statistical data sets—GOV.UK (2016)
Flexiblepower Alliance Network: The PowerMatcher Suite. http://flexiblepower.github.io/. Accessed 28 Apr 2017
Luo Y, Itaya S, Nakamura S, Davis P (2014) Autonomous cooperative energy trading between Prosumers for Microgrid systems, 693–696
Murkin J, Chitchyan R, Byrne A (2016) Enabling peer-to-peer electricity trading. In: 4th International Conference on ICT for Sustainability
Power Ledger-A New Decentralised Energy Marketplace—Where Power meets Blockchain (2016)
Wang Y et al (2014) A game-theoretic approach to energy trading in the smart grid. IEEE Trans Smart Grid 5(3):1439–1450
Yaagoubi N, Mouftah HT (2015a) Energy trading in the smart grid: a game theoretic approach. In: IEEE SEGE, pp 1–6
Yaagoubi N, Mouftah HT (2015b) A distributed game theoretic approach to energy trading in the smart grid. In: IEEE EPEC, pp 203–208

Part IV
Software Tools and Environmental Databases

A Literature Survey of Information Systems Facilitating the Identification of Industrial Symbiosis

Guido van Capelleveen, Chintan Amrit and Devrim Murat Yazan

Abstract Industrial Symbiosis (IS) is an emerging business tool that is used by practitioners to engage cooperation among industries to reuse waste streams. The key to reveal IS opportunities for organizations is both connecting the supply and demand of various industries and providing technical knowledge on the IS implementation. This process is increasingly supported by information systems which act as a facilitator of communication and distributor of knowledge. However, we lack understanding of a describing role of each type of information system within the process of IS identification. IS literature could benefit from a clear overview of (i) the characteristics of these different information systems, (ii) the role of support these systems provide, and (iii) the technologies used to enable such identification. This paper analyzes the current state of literature that addresses information systems that facilitate IS identification and studies these systems using these three pillars. Our study contributes by providing a classification framework of information systems that facilitate industrial symbiosis identification and reveals three research directions to progress IS identification tools, namely (i) software product and service development (ii) data integration, and (iii) adoption of intelligent learning.

1 Introduction

The reduction of waste emissions, primary resource and energy use in resource-intensive industries contributes to the development of a sustainable environment (European Environmental Agency 2016). One of the attempts to reduce waste, emissions and resource use is industrial symbiosis. Industrial symbiosis (IS) is a cooperation among industries wherein waste streams as a secondary output of an industry, are utilized as (part of) resources for the production processes

G. van Capelleveen (✉) · C. Amrit · D.M. Yazan
Department of Industrial Engineering and Business Information Systems,
University of Twente, PO box 217, 7500 AE Enschede, The Netherlands
e-mail: g.c.vancapelleveen@utwente.nl

© Springer International Publishing AG 2018
B. Otjacques et al. (eds.), *From Science to Society*, Progress in IS,
DOI 10.1007/978-3-319-65687-8_14

of other industries (Chertow 2000). The IS methodology is gaining popularity as a means to improve sustainable production because of its ability to provide economic benefits while simultaneously providing ecological benefits as well as social benefits (Chertow 2007). While most studies focus on evolutionary models (Boons et al. 2016), less attention is devoted towards the use of information systems.

IS literature still lacks a clear overview of the different type of systems and the functional support these systems can provide. In particular, there is a gap of knowledge in the understanding of the role of information systems as a tool to support industrial symbiosis development. This study provides an overview with a focus on the systems facilitating the identification of IS. We contribute by providing a conceptual framework to practitioners and policy makers that outlines the course of actions for information systems that facilitate IS identification. Moreover, this review reveals research directions for the IS community to consider by suggesting advancements to be made in order to strengthen community engagement and enrich data use for computerized exploration.

2   Research Design

We analyze the Industrial Symbiosis (IS) literature from 2000 to 2016. We have selected the time-frame based on the publication date of the well cited paper (Chertow 2000), which can be seen as one of the early foundations on the concepts of industrial symbiosis. To capture the scarcely published research on IS tools we take an inter-disciplinary approach in retrieving articles from a variety of journals and conferences, as information system literature is scattered over a multi-fold of disciplinary fields ranging from ecological ecosystems to computer sciences.

We initially conducted a Scopus (Elsevier 2017) search for bibliographic scientific material. Our queries search for a set of keywords in the title, abstract and keywords of articles. The keywords reflected the concepts of "industrial symbiosis" and "information systems" and were applied in various combinations between the concepts. The keywords were "industrial symbiosis, industrial ecology, eco-industrial park, synergy, by-product exchange, recycling network, waste exchange", and "information system, ICT, tool, decision support, intelligence, expert system, identification, assessment, mapping". To prevent missing articles due to the use of inconstant terminology practiced over different disciplines, we performed a first degree backward and forward analysis based on the citations of the relevant publications (snowballing technique). We use the following selection criteria to filter relevant IS articles that we included in our review:

- The article mentions, or presents an IS identification tool that is primarily focused to reveal options, industries or individual organizations in support of waste exchange or industrial symbiosis network development.
- The tool is considered an information system, and thus utilizes various ICT techniques.

- The article provides extensive information on the implementation of the tool or elaborates on the techniques that are essential and specific to this type of information systems.

Then, we use the selected literature to conceptualize the IS systems into a matrix using the methodology of Webster and Watson (2002) to further develop an understanding of the key concepts of the information systems that stimulate IS facilitation.

3 Industrial Symbiosis Tools

This section describes our typology consisting of six concepts (see Table 1), created from the identified literature, and summarizes advancements of each of the areas.

Table 1 Literature classification framework of information systems that facilitate industrial symbiosis identification (Note that Grant et al. 2010; Veiga and Magrini 2009; Massard and Erkman 2007, 2009 discuss more than one concept)

Type	Information system	Type of support	Characteristics	Common technology and techniques	References
A	Open online waste markets	Passive facilitation of waste transactions	A web-based information system that provides an open business-to-business sales platform service for waste materials. The platforms do not or -to a minimal extend- interfere or coordinate transactions. Most systems form an open market and information is visible to the outside world. Typically, this type of market is mostly driven by governmental agencies or individuals from an idealistic view without a sustainable profitable business model	Web-based e-commerce platforms, rule-based matchmaking algorithms, ontology engineering	(Cecelja et al. 2015; Chen et al. 2006; Dhanorkar et al. 2015; Dietrich et al. 2014; Grant et al. 2010; Große et al. 2016; Hein et al. 2015; Hickey et al. 2014; Raafat et al. 2013; Trokanas et al. 2015)

(continued)

Table 1 (continued)

Type	Information system	Type of support	Characteristics	Common technology and techniques	References
B	Facilitated synergy identification systems	Active facilitation of waste transactions	An information system used to support private community-based business-to-business sales of waste materials which in many cases are facilitated through an intermediary. These type of systems are used during or after identifications workshops and tend to be more structured and coordinated by the intermediary. Typically, these markets are initiated by consortia, industrial parks or third party facilitator and are more driven upon profitable results	Rule-based matchmaking algorithms, national input-output tables, and life-cycle assessment databases	(Alvarez and Ruiz-Puente 2016; Baas and Hjelm 2015; Clayton et al. 2002; Cutaia et al. 2014, 2015; Grant et al. 2010; Massard and Erkman 2007;, 2009; Mirata 2004; Raabe et al. 2017; Song et al. 2017; Sterr and Ott 2004; Veiga and Magrini 2009; Zhang et al. 2017)
C	Industry sector synergy identification	Profiling of waste production and use per industry	A system approach that examines synergies between industry sectors rather than between factories. Using national waste statistics per industry sector it is possible to detect the abundant waste inputs and outputs. Using this information and the help of manufacturing process information it is possible to determine waste flows that can potentially be converted into synergies	National waste input-output tables	(Chen 2015; Horvath 2016)

(continued)

Table 1 (continued)

Type	Information system	Type of support	Characteristics	Common technology and techniques	References
D	Social network platforms and social network communities	Mixed: (a) Building social relations (b) Knowledge exchange	The role of Social Networks (SN) in sharing knowledge and information on IS experiences and opportunities among industries. The tools unfold either as independent SN platforms or as a formation of groups on existing SN platforms	Social networks, and social network integration	(Ghali et al. 2016)
E	Industrial symbiosis knowledge repositories	Knowledge exchange	Knowledge systems that contain a collection of historical industrial symbiosis examples or theoretical literature references to potential IS cases. The systems enable collaborative knowledge creation by providing a platform to share and discuss synergy implementation experiences	Content Management systems (e.g. Wikis), and collaborative web-based spreadsheets	(Davis et al. 2010; ISDATA 2015; Veiga and Magrini 2009; Yu et al. 2014)
F	Region identification system for industrial symbiosis	Urban planning and policy making	Geographical information systems (GIS) that estimate the potential of areas for the application of industrial symbiosis development	GIS visualizations, Geographical-data based scoring algorithms, fuzzy rule based expert systems, and simulation software	(Aid et al. 2015; Dou et al. 2016; Massard and Erkman 2007, 2009; Ruiz et al. 2012; Togawa et al. 2016)

3.1 Open Online Waste Markets

Open online waste markets or waste e-marketplaces are web-based platforms for business to trade industrial residues. Three key characteristics are observed in this category of information systems. Firstly, the markets provide an electronic service to connect organizations' supply and demand, such as in other traditional e-marketplaces. Secondly, the marketplaces are open to any industry and items that are publicly listed. Finally, the marketplaces act as an independent service without active involvement of IS facilitators. The diversification of systems generally result from the type of commodities that are transferred, the geographical areas covered and the type of actors who initiate these systems. The U.S. Environmental Protection Agency (EPA) defined these online material and waste 'exchanges' as *"markets for buying and selling reusable and recyclable commodities. Some are physical warehouses that advertise available commodities through printed catalogs, while others are simply web sites that connect buyers and sellers"* (Environment, Health and Safety 2017). The labeling of these symbiosis as exchanges is contested by the IS definition provided by (Lombardi and Laybourn 2012), who argue these could be better referred to as business transactions. These online waste markets also predominantly select a business model considering waste as transactional items rather than for forming collaborative relations in order to continuously reuse multiple wastes among industries.

Most electronic waste markets originate as a result of promoting the concept of waste transfers between industries in Europe in the beginning of 1970 (U.S. EPA 1994). During this period, clearinghouses acted as intermediaries to list the wastes of industries. However, the success of the Internet in the 90 s increased the ease of communication linking regional markets together and resulted in the online material and waste markets as we know them today. The success of e-marketplaces is mainly attributed to the involvement of industrial actors in establishing a critical mass of users and creating sufficient supply and demand (Evans and Schmalensee 2010). Only a few of these IS markets attract this critical mass, whereas the majority fail to become largely populated and as a consequence do not take off (see e.g. (Environment, Health and Safety 2017)). Moreover, the established markets struggle to sustain their business. This is, among other reasons, because the market existence is based on a continuous flow of supply and demand, which requires the participating organizations to repeatedly invest in the identification of new potential waste transactions or frequently attract new organizations to join. Although many of these systems struggle to become economically viable and find it difficult to engage industries, there is potential that is recognized to gain both economic and ecologic benefits (e.g. in a case of construction and demolition waste (Chen et al. 2006)). The challenge often is to bootstrap such systems, such as in The Resource eXchange Platform (TRXP) (Dietrich et al. 2014; Hickey et al. 2014), a resource 'exchange' web-platform facilitating industrial symbiosis by building a network of organizations in Europe that could reuse or disassemble industrial streams of ICT equipment. The researchers (Dietrich et al. 2014) show that the open platform is technically feasible and could extend reuse streams by means of industrial symbiosis. However, the promoters also demonstrate with scaling their concept of re-usability that economic validity remains difficult, and that regulatory constraints affect the success of such networks.

Research that investigates a viable example of an online waste platform operating in Minnesota, United States (Dhanorkar et al. 2015), identify patterns in the successful transactions and relate positive effects back to the presence of rich re-purposing information and product information. Providing sellers access to re-purposing alternatives positively affected their network commitment.

Furthermore, the rich product information and transaction process explanation used to inform buyers is also found to positively affect successful transactions by reducing the buyers' uncertainty. Interestingly, the analysis reveals that items that require longer term contracting are underrepresented in the successful transactions and are suggested to benefit from more intensive facilitation that supports the contracting, reduces the uncertainty and builds trust (Dhanorkar et al. 2015).

From a technical perspective, ontological engineering techniques studied by (Raafat et al. 2013) demonstrate the ability to enhance and structure waste item information in e-marketplaces, such as in the e-symbiosis system (Cecelja et al. 2015). These ontologies enable the use of systematic matching in online systems that support waste transactions (Trokanas et al. 2015). The design of techniques that should match process waste to primary inputs raises some challenges. For example, an input-output matching approach, which basically matches supply and demand items based on a classification code, would identify only the resources that could directly be matched, while (Hein et al. 2015) points out also to consider the process of substitution. This suggests to provide relationships between wastes and the potential resulting materials from processing, in matching wastes to inputs. Connecting industries through web-based interfaces gains popularity while slowly also becomes an object for investigation to energy based symbiosis (Große et al. 2016).

3.2 Facilitated Synergy Identification Systems

Our second information systems concept is classified as facilitated synergy identification systems. Although technically these type of systems share to a large extent similar technologies with open online waste markets, from organizational perspective these systems are quite distinct. The major difference is found in how the systems are being used by clusters and facilitators to explore IS; aligning this group of systems with the facilitated industrial symbiosis approach (Paquin and Howard-Grenville 2012), or in few cases to planned conditions in eco-industrial parks or IS networks (Boons et al. 2016; Chertow 2007). Furthermore, these systems typically adopt more coordinated forms of waste identifications, few are publicly accessible and tend to have a larger impact in creating sustainable long term oriented symbiosis from waste streams than most open online waste markets (e.g. (Sterr and Ott 2004)).

The initial waste transactions often created using these type of IS tools are generally the result of organized IS identification workshops, within eco-industrial clusters or among regional participants from different industrial sectors (Mirata 2004; Paquin and Howard-Grenville 2012; van Beers et al. 2007). During the workshops synergy identification tools can assist providing in identification support and exploration of IS opportunities. This type of support varies from input-output matching to the review of substitutes, affective regulations or the assessment of financial and environmental impacts (Grant et al. 2010). In addition, after the initiation of such networks, the systems can potentially serve as a communication hub for an internal waste market. It also has been suggested that academic and business conferences can play such a role by disseminating ideas and helping with the identification of sustainable transactions (Baas and Hjelm 2015).

Remarkably, many studies acknowledge the support given by synergy identification systems. However, few reach the desired long-term impact, and remain operative over a period of time (Grant et al. 2010). This is observed in the WasteX system in Jamaica which demonstrates the feasibility of a web-based waste 'exchange' system developed for developing economies. The project struggled for a while with industrial adoption due to a lack of critical mass of users (Clayton et al. 2002), and later became inoperative (Grant et al. 2010). The study of (Grant et al. 2010) reveals that the majority of systems have not been designed with a commercial prospective in mind. On the contrary, the system development concentrates on the functionality; which is provided as one of the main reasons why many systems have failed to sustain. This can be related back to the fact that tools often do not consider the need to enhance social participation; focusing more on determining technical opportunities rather than building human relationships. Furthermore, positive effectors considered to succeed sustainable markets are the industry adoption of standardized waste taxonomy and the presence of a key number of organizations in order to bootstrap the newly created market.

Examples of IS tools resulting in fruitful and sustainable collaborations include the Institute of Eco-Industrial Analysis (IUWA) Waste manager implemented applied in the Rhine-Necklar region in Germany (Chen et al. 2006), and the information management tools used in the National Industrial Symbiosis Program (NISP) in the United Kingdom (Mirata 2004). The study of (Sterr and Ott 2004) shows that such identification tools can promote the potential for symbiotic cooperation at a wider regional level in order to establish more connected industrial waste reuse networks. Nonetheless, enlarging regional size presents new challenges including higher organizational costs, building trust and coordination and requires more practical ways of information exchange to detect appropriate partners (Mirata 2004; Sterr and Ott 2004) shows a similar tool deployed in the National Industrial Symbiosis Program (NISP) in the United Kingdom. A material flow database in combination with a decision support tool is used to reveal feasible industrial synergies and to evaluate the environmental effect of IS actions. Here, the important role of coordinating bodies is to provide support for a whole range of activities, including processing, logistics, capacity management of regional parties, technology alternatives, environmentally preferable practices, market dynamics and regulatory issues (Mirata 2004). These ideas emerged into the SynerGIE system practiced by the facilitators in the NISP program, and is now applied in many other countries (Ghali et al. 2016; Lombardi and Laybourn 2015).

Recent studies (Cutaia et al. 2015; van Beers et al. 2007; Veiga and Magrini 2009) continue to experiment with tool support in facilitated form to create IS relations. They investigate not only the effectiveness of exploring new synergies, but also address numerous forms of synergy assessment support that may be given with these systems, for example, in a project in the Kwinana and Gladstone mineral industry in Australia (van Beers et al. 2007). The project aims at developing practical assistance to the industries with tools that could identify and develop new synergies with sustainable outcomes. Furthermore, we observe attempts at creating and adopting packaged software solutions (commercial of-the-shelf) in facilitated IS research. An example of this is the Facility Synergy Tool (FaST) developed by the U.S. Environmental Protection Agency (EPA) that was used to identify IS opportunities in the eco-industrial-cluster in Rio de Janeiro (Veiga and Magrini 2009). The integration of identification and assessment components was part of these tools. These elements are combined into an overall platform in order to provide a toolbox that can be used by practitioners. Similarly, the IS platform developed in Catania, Italy, reports one platform combined by geographical information, environmental regulations, best IS practices and to a certain extent life cycle assessment technology (Cutaia et al. 2015, 2014). Showing the economic viability to organizations as a component of a platform aimed at IS identification is also argued as one of major facilitators (Raabe et al. 2017). A data oriented approach would both help to reveal new industrial symbiosis opportunities as well to assess the economic benefits that could demonstrate the IS viability (Song et al. 2017).

3.3 Industry Sector Synergy Identification

Another interesting concept applied in information systems is to examine synergies among industries rather than between firms which is demonstrated in a case study in Taiwan (Chen 2015). With the help of input-output tables, opportunities to use by-products were identified for each industry type, and used as suggestions for organizations that share common features and production processes. Following this holistic view to link industry sectors, a hypothetical decision support system is suggested that could review the duties and responsibilities of authorities in IS creation and then support them through the analysis of production industries to detect open and fixed material loops (Horvath 2016). This type of information system would not only support direct identification, but also help in developing new environmental policies that stimulate industries to adopt industrial symbiosis.

3.4 Social Network Synergy Identification

A fairly new development considers the potential of social network platforms to identify and develop groups for actors who share IS interests. Social media can be a helpful tool to distribute information, build relationships, share experiences and coordinate communities. The suggested web-based platform, that is based on social network content, could serve an IS community to forge new relationships and share experiences (Ghali et al. 2016). Although organizations may be hesitant to share information extensively using these networks, they might be effective in introducing people who work on similar topics, as observed in industry related LinkedIn groups (2017).

3.5   Industrial Symbiosis Knowledge Repositories

IS knowledge repositories can be helpful tools for organizations to reveal new potential synergies. Not only do they enable organizations with repositories to share their IS experiences, but also provide them with a gateway to a variety of structured and unstructured information on industrial symbiosis. Nowadays, a few of the IS knowledge repositories started to develop web interfaces that authorize access to databases, e.g. life cycle inventories, resource substitution data, IS technologies, regulations, and services that support taxonomy translation.

A good example of an IS platform is the Industrial Symbiosis DATA repository (ISDATA 2015). The platform collects IS case studies, cross analyzes material properties and maps open data sets related to the identification of IS. Many other data collections, both developed by businesses and governmental agencies, also act as reference guides in IS identification, such as Nordregio IS cases (Nordregio 2016), the Industrie et Synergies Inter-Sectorielles database, and the collection of NISP case studies (Enipedia 2015). Another type of repositories are IS Wiki's, which are community platforms that enable collaborative data collection and moderation, in which knowledge of implemented IS case examples are shared, e.g. Enipedia (Davis et al. 2010; Yu et al. 2014). What we observe in many tool implementations is that due to the lack of economic incentives, it is difficult to continuously develop and maintain such information sources. This phenomena is also more widely observed in the field of knowledge systems where this lack of active contribution is explained clearly by the social dilemmas of knowledge sharing (Cabrera and Cabrera 2002; Wang and Noe 2010). The dilemma refers to the situation that sharing insights come at a cost of an individual while there usually is less to gain from the sharing act itself. On the other hand, commercial organizations collect and sell such information, but consider a business model to access and use such data. Moreover, data are affected by legal constrains and often require confidentiality; hence, challenges to the design of information use in such platforms.

However, many researchers have pointed out using environmental data sets can add value to IS identification systems (Cutaia et al. 2015; Davis et al. 2009; Mattila et al. 2012; Sterr and Ott 2004). Incorporation of these external information sources into facilitation tools can enhance the exploration and assessment model. Analyzing the IS case data would enable the creation of expert systems that could recommend based on historical cases. However, the ability to link all such information generally requires common, unambiguous terminology or a service for taxonomy translation (ISDATA 2015; Sander et al. 2008). Common resource and waste classification standards, such as the widely used European Waste Catalogue (EWC) (European Commission 2000), are not ideal for this purpose, as there is overlap of classifications, ambiguity in classifications and, in some cases, an inadequate level of detail not sufficient to IS exploration (Sander et al. 2008).

3.6 Industrial Symbiosis Region Identification

Our final category considers systems that aim to support the strategic location of industrial symbiosis investments. We conceptualize these tools as IS region identification systems. The main function of such systems is to identify regions conducive for industrial symbiotic development, mostly with geographical interface elements, such that industrial park planners and facilitators of IS can optimize their investment of time and resources. For example, in the IS project situated in Geneva where a geographical analysis tool is being developed that visualizes the economic activity of the Swiss canton (Massard and Erkman 2007, 2009). The key of the system is its function to locate the regions that have a high economic viability for IS implementations. The approach of region identification is in line with IS facilitators who suggest that current industrial sites with existing cooperation provide the best prospects for identifying new industrial symbiosis opportunities. The design of a symbiosis suitability index demonstrates this ability to identify IS regions at the South Humber Bank (Jensen et al. 2012).

Visualizing the economic activity as an assessment of IS viability of a region requires one to connect to a variety of data sources. In general, aggregated data reflecting regional economic, infrastructural or industrial characteristics are most useful in determining the applicability of IS. Characteristic examples include land destination, the projection of waste production, network-density of infrastructures, urban development, diversity of industry, availability of nearby facilities such as power plants, boreholes and waste facilities (Aid et al. 2015; Jensen et al. 2012; Ruiz et al. 2012). Typically, with such a variety of inputs, the decision process involves a multi-criteria evaluation. Hence scoring algorithms or fuzzy rule based expert systems are common approaches that provide such spatial decision support. One method of conducting multi-criteria evaluation is to create weighted normalized evaluations using Analytic Hierarchy Process (AHP), as the one performed in the Cantabria region in northern Spain (Ruiz et al. 2012). The study of (Aid et al. 2015) from Sweden, shows a heuristic tool named Looplocal. The tool assesses IS potential through analyzing resource flow data of regions having the IS knowledge on inputs and outputs derived from a LCA database and a collection of IS case studies. Though there is some interest from the industry for such systems, additional evaluation of relevance is still emphasized (Aid et al. 2015).

Region identification could also be deployed to determine the best location for specific IS cases. For example, this is seen in the spatial planning of district heating systems (Togawa et al. 2016). District heating focuses on the recovery of waste heat from municipal solid waste incinerators and low temperature industrial waste heat utilized for district heating in urban areas. The urban sustainable energy planners in Japan experiment with geographical information systems that analyze the waste heat potential of their regions (Dou et al. 2016). These systems evaluate the usability, feasibility and efficiency of a district heat system quantitatively by calculating CO_2 reductions and through conducting cost benefit analyses of multiple scenarios. Note that most of these regional identification system examples from literature are fairly conceptual and not yet widely adopted by industry or governmental planners, presumably for reasons of economic viability.

4   IT Challenges for Industrial Symbiosis Tools

From an information systems perspective we can derive three key observations that provide directions for the further development of identification tools for industrial symbiosis. The first direction is to move from custom made software (observed in (Grant et al. 2010)), towards mass product software. We believe that the application context (e.g. the parks/regions, industrial sectors, and active regulations) is not that different and hardly affects the type of functions required. This suggests reusing software and moving towards an IS tool package that can be deployed in many situations. Often such a transition of software is associated with the terminology of product software (Xu and Brinkkemper 2007). Secondly, we consider the integration of documented IS cases, linked data and IS tools into a platform. Furthermore, a combination of services in a central environment can streamline work processes, and a utilization of various data sources may help reveal new insights that lead to new type of services. The key requirement for this is the adoption of service oriented architectures into the design of the decision support systems (Demirkan and Delen 2013). A key software requirement to data providers, such as LCA databases, is to provide open accessible interfaces to enable adoption of data services within the tool package that can enhance or enable the use of IS explorations. The advantage of such an approach is that it can lead to the development of services for different devices and multiple platforms (digital mesh). A third research direction for tools is the tailoring of intelligent learning algorithms to the market of industrial symbiosis. Because of the increase in expertise and growing availability of data intelligent learning is slowly becoming more relevant. A particular set of machine learning techniques that in our view apply to IS markets are recommender algorithms (Ekstrand et al. 2011). Popular recommendation techniques that could be interesting to evaluate in the context of IS identification are association rule mining, case-based reasoning, collaborative filtering, knowledge-based recommendation, and rule-based recommendation.

5   Conclusions

Overall, this article reviewed 16 years of literature (2000–2016) on information systems used to facilitate the identification of industrial symbiosis. We provided a framework with six identified concepts to better understand IS identification tools, namely; open online waste markets, facilitated synergy identification systems, industry sector synergy identification, social network platforms, IS knowledge repositories and regional identification systems. In general, many of the identification tools presented in literature report difficulties with wide industry adoption, but also recognize the benefits that potentially can be derived. The key challenge for IS tools is demonstrating those economic, social and ecological benefits of potential IS opportunities to industries, while simultaneously preserving data confidentiality, conforming to regulations and assuring the stakeholders against risks resulting from the IS implementation. Along with the growth in IS experience reports and IS data sources, future research may focus on the development of product and service oriented IS software, the integration of data and tools into a tool package and the adoption of intelligent learning techniques.

Acknowledgements This research is funded by European Union's Horizon 2020 program under grant agreement No. 680843.

References

Aid G, Brandt N, Lysenkova M, Smedberg N (2015) Looplocal a heuristic visualization tool to support the strategic facilitation of industrial symbiosis. J Clean Prod 98:328–335

Alvarez R, Ruiz-Puente C (2016) Development of the tool symbiosys to support the transition towards a circular economy based on industrial symbiosis strategies. Waste Biomass Valorization, pp 1–10

Baas L, Hjelm O (2015) Support your future today: enhancing sustainable transitions by experimenting at academic conferences. J Clean Prod 98:1–7

Boons F, Chertow M, Park J, Spekkink W, Shi H (2016) Industrial symbiosis dynamics and the problem of equivalence: proposal for a comparative framework. J Ind Ecol

Cabrera A, Cabrera EF (2002) Knowledge-sharing dilemmas. Organ Stud 23(5):687–710

Cecelja F, Raafat T, Trokanas N, Innes S, Smith M, Yang A, Zorgios Y, Korkofygas A, Kokossis A (2015) e-Symbiosis: technology-enabled support for industrial symbiosis targeting small and medium enterprises and innovation. J Clean Prod 98:336–352

Chen PC, Ma HW (2015) Using an industrial waste account to facilitate national level industrial symbioses by uncovering the waste exchange potential. J Ind Ecol 19(6):950–962

Chen Z, Li H, Kong SC, Hong J, Xu Q (2006) E-commerce system simulation for construction and demolition waste exchange. Autom Constr 15(6):706–718

Chertow MR (2000) Industrial symbiosis: literature and taxonomy. Annu Rev Energy Env 25(1):313–337

Chertow MR (2007) Uncovering industrial symbiosis. J Ind Ecol 11(1):11–30

Clayton A, Muirhead J, Reichgelt H (2002) Enabling industrial symbiosis through a web-based waste exchange. Gr Manag Int 40:93–107

Cutaia L, Luciano A, Barberio G, Sbaffoni S, Mancuso E, Scagliarino C, La Monica M (2015) The experience of the first industrial symbiosis platform in italy. Environ Eng Manag J 14(7):1521–1533

Cutaia L, Morabito R, Barberio G, Mancuso E, Brunori C, Spezzano P, Mione A, Mungiguerra C, Li Rosi O, Cappello F (2014) The project for the implementation of the industrial symbiosis platform in sicily: the progress after the first year of operation. Springer International Publishing, Cham, pp 205–214

Davis C, Nikoli I, Dijkema GPJ (2009) Integration of life cycle assessment into agent-based modeling. J Ind Ecol 13(2):306–325

Davis C, Nikolic I, Dijkema GP (2010) Industrial ecology 2.0. J Ind Ecol 14(5):707–726

Demirkan H, Delen D (2013) Leveraging the capabilities of service-oriented decision support systems: putting analytics and big data in cloud. Decis Support Syst 55(1):412–421

Dhanorkar S, Donohue K, Linderman K (2015) Repurposing materials and waste through online exchanges: overcoming the last hurdle. Prod Oper Manage 24(9):1473–1493

Dietrich J, Becker F, Nittka T, Wabbels M, Modoran D, Kast G, Williams I, Curran A, den Boer E, Kopacek B, et al (2014) Extending product lifetimes: a reuse network for ict hardware, vol 167, pp 123–135

Dou Y, Togawa T, Dong L, Fujii M, Ohnishi S, Tanikawa H, Fujita T (2016) Innovative planning and evaluation system for district heating using waste heat considering spatial configuration: a case in fukushima, Japan. Resour Conserv Recycl

Ekstrand MD, Riedl JT, Konstan JA (2011) Collaborative filtering recommender systems. Found Trends Hum Comput Interact 4(2):81–173

Elsevier (2017) Scopus document search. https://www.scopus.com/. Accessed 01 Mar 2017

Enipedia (2015) Case studies. http://enipedia.tudelft.nl/wiki/Industrial_Symbiosis_Data_Sources \#CASE_STUDIES. Accessed 16 Feb 2017

Environment, Health and Safety (2017) List of U.S. state-specific waste exchanges. http://www.ehso.com/wastexchg.php. Accessed 16 Feb 2017

European Commission (05 2000) Commission decision on the european list of waste (com 2000/532/ec). Technical report, European Commission

European Environmental Agency (10 2016) More from less material resource efficiency in europe. eea report no 10/2016. Technical report, European Environmental Agency

Evans DS, Schmalensee R (2010) Failure to launch: critical mass in platform businesses. Rev of Netw Econ 9(4)

Ghali MR, Frayret JM, Robert JM (2016) Green social networking: concept and potential applications to initiate industrial synergies. J Clean Prod 115:23–35

Grant GB, Seager TP, Massard G, Nies L (2010) Information and communication technology for industrial symbiosis. J Ind Ecol 14(5):740–753

Große J, Matusevicus A, Wohlgemuth V (2016) Prototypische umsetzung einer webanwendung zur beurteilung von stoff- und energieströmen am standort berlin-schöneweeide. In: Environmental informatics—stability, continuity, innovation: current trends and future perspectives based on 30 years of history, pp 99–105

Hein AM, Jankovic M, Farel R, Sam LI, Yannou B (2015) Modeling industrial symbiosis using design structure matrices. In: 17th international dependency and structure modeling conference, DSM 2015

Hickey S, Fitzpatrick C, Maher P, Ospina J, Schischke K, Beigl P, Vidorreta I, Yang M, Williams I (May 2014) A case study of the d4r laptop, vol 167, pp 101–108

Horvath G (2016) A framework for an industrial ecological decision support system to foster partnerships between businesses and governments for sustainable development. J Clean Prod 114:214–223

ISDATA (2015) The industrial symbiosis data repository. http://isdata.org/. Accessed 31 Jan 2017

Jensen PD, Basson L, Hellawell EE, Leach M (2012) habitat suitability index mapping for industrial symbiosis planning. J Ind Ecol 16(1):38–50

LinkedIn (2017) Industrial symbiosis. https://www.linkedin.com/groups/1845383. Accessed 02 Mar 2017

Lombardi DR, Laybourn P (2012) Redefining industrial symbiosis. J Ind Ecol 16(1):28–37

Lombardi D, Laybourn P (2014) National industrial symbiosis programme (nisp): Connecting industry, creating opportunity. In: 2015 ENEA, 2012(2013), p 22

Massard G, Erkman S (2007) A regional industrial symbiosis methodology and its implementation in geneva, switzerland. In: 3rd international conference on life cycle management, University of Zurich, Citeseer, Irchel

Massard G, Erkman S (2009) A web-gis tool for industrial symbiosis: Preliminary results and perspectives. In: 23rd international conference on informatics and environmental protection

Mattila T, Lehtoranta S, Sokka L, Melanen M, Nissinen A (2012) Methodological aspects of applying life cycle assessment to industrial symbioses. J Ind Ecol 16(1):51–60

Mirata M (2004) Experiences from early stages of a national industrial symbiosis programme in the uk: determinants and coordination challenges. J Clean Prod 12(810):967–983

Nordregio (2016) 50 industrial symbiosis case studies. http://www.nordregio.se/50cases. Accessed 16 Feb 2017

Paquin RL, Howard-Grenville J (2012) The evolution of facilitated industrial symbiosis. J Ind Ecol 16(1):83–93

Raabe B, Low JSC, Juraschek M, Herrmann C, Tjandra TB, Ng YT, Kurle D, Cerdas F, Lueckenga J, Yeo Z, Tan YS (2017) Collaboration platform for enabling industrial symbiosis: Application of the by-product exchange network model. Procedia CIRP 61:263–268

Raafat T, Trokanas N, Cecelja F, Bimi X (2013) An ontological approach towards enabling processing technologies participation in industrial symbiosis. Comput Chem Eng 59:33–46

Ruiz M, Romero E, Prez M, Fernndez I (2012) Development and application of a multi-criteria spatial decision support system for planning sustainable industrial areas in northern spain. Autom Constr 22:320–333

Sander K, Schilling S, Lskow H, Gonser J, Schwedtje A, Kchen V (11 2008) Review of the european list of waste. Technical report, kopol GmbH and ARGUS GmbH

Song B, Yeo Z, Kohls P, Herrmann C (2017) Industrial symbiosis: Exploring big-data approach for waste stream discovery. Procedia CIRP 61:353–358

Sterr T, Ott T (2004) The industrial region as a promising unit for eco-industrial development reflections, practical experience and establishment of innovative instruments to support industrial ecology. J Clean Prod 12(810):947–965

Togawa T, Fujita T, Dong L, Ohnishi S, Fujii M (2016) Integrating (GIS) databases and (ICT) applications for the design of energy circulation systems. J Clean Prod 114:224–232

Trokanas N, Cecelja F, Raafat T (2015) Semantic approach for pre-assessment of environmental indicators in industrial symbiosis. J Clean Prod 96:349–361

U.S. EPA (1994) Review of industrial waste exchanges. Technical report, U.S. Environmental Protection Agency

van Beers D, Corder G, Bossilkov A, van Berkel R (2007) Regional synergies in the australian minerals industry: Case-studies and enabling tools. Miner Eng 20(9):830–841

Veiga LBE, Magrini A (2009) Eco-industrial park development in rio de janeiro, brazil: a tool for sustainable development. J Clean Prod 17(7):653–661

Wang S, Noe RA (2010) Knowledge sharing: A review and directions for future research. Hum Resour Manage Rev 20(2):115–131

Webster J, Watson RT (2002) Analyzing the past to prepare for the future: writing a literature review. MIS Quarterly 26(2):xiii–xxiii

Xu L, Brinkkemper S (2007) Concepts of product software. Eur J Inf Syst 16(5):531–541

Yu C, Davis C, Dijkema GP (2014) Understanding the evolution of industrial symbiosis research. J Ind Ecol 18(2):280–293

Zhang C, Romagnoli A, Zhou L, Kraft M (2017) Knowledge management of eco-industrial park for efficient energy utilization through ontology-based approach. Appl Energy

Using Twitter for Geolocation Purposes During the Hanse Sail 2016 in Rostock

Ferdinand Vettermann, Christian Seip and Ralf Bill

Abstract This work is embedded in the project KOGGE, which is related to smaller urban water bodies in the city of Rostock. Public participation in water management issues is one of the goals. Thus, twitter messages are an interesting data source, especially for spatially related issues. We developed a strategy to analyze and match twitter messages to locations/events in a smaller city, like Rostock, with the help of a gazetteer. To evaluate our strategy, we used the Hanse Sail 2016 as a big event with nearly one million visitors. We chose such a big event to present the use of such an analysis and to get a bigger database in a shorter timeframe. From 10.08.2016–16.08.2016 we were able to collect more than 6.2 Mio. tweets written in German language. With our gazetteer-matching method we are able to collect a sufficient amount of geolocated tweets. We identify places of particular significance, like the big entertainment stages in the city harbor of Rostock. Finally, we want to develop a tool, which is able to support decisions in urban planning with the usage of social media analysis.

Keywords Twitter · Volunteered geographic information · Hanse sail · Public participation · Gazetteer matching · Geolocation extraction

1 Introduction

Within the BMBF (Federal Ministry of Education and Science) funded project KOGGE ("Kommunale Gewässer Gemeinschaftlich Entwickeln") a key issue is to integrate the public in urban planning processes. Project partners are the University

F. Vettermann (✉) · C. Seip · R. Bill
Professorship of Geodesy and Geoinformatics, University of Rostock, Rostock, Germany
e-mail: ferdinand.vettermann@uni-rostock.de

C. Seip
e-mail: christian.seip@bkg.bund.de

R. Bill
e-mail: ralf.bill@uni-rostock.de

© Springer International Publishing AG 2018   171
B. Otjacques et al. (eds.), *From Science to Society*, Progress in IS,
DOI 10.1007/978-3-319-65687-8_15

of Rostock (professorship of water management, professorship of hydrology, professorship of geodesy und geoinformatics), the "EURAWASSER Nord GmbH", the water and soil association "Untere Warnow-Küste" and "biota – Institut für ökologische Forschung und Planung GmbH". They cooperate with the city of Rostock, senate administration for construction and environment, the Warnow-water and wastewater association, the department of agriculture and environment of central Mecklenburg-Western Pomerania and the federal office for environment, nature conservation and geology Mecklenburg-Western Pomerania. Thus, different stakeholders of water management and urban planning in the area of Rostock participate in this project.

Social networks provide a growing research field. They address the interest of the people to know, what others are doing and with whom. Location Based Social Networks (LBSN) give the user the possibility to tag posts with location information (Traynor and Curran 2013). The location element in social networks spread over the last years with the GPS sensors in mobile devices (Goodchild 2007). These locations are of special interest in recent research, because they give the possibility to investigate movement of people (Frias-Martinez et al. 2012), identify points of interest and the sentiment of statements in different city quarters (Agarwal et al. 2011) or to locate and track floods or other hazardous events (Fuchs et al. 2013; Li et al. 2012). The location information can be incooperated as a place added by the user itself, with the help of a GPS coordinate, it is predictable from IP-addresses or WLAN hotspots or it is written directly (or indirectly) into the message text as a Location Indicative Word (LIW) (Grahem et al. 2013).

2   Common Issues When Locating Tweets from Text

It is still a common problem to extract location information from social network messages, especially to identify objects in urban landscapes in high resolution. A difficult task is to extract useful information associated with location from the highly unstructured twitter messages (Grahem et al. 2013). Especially in Germany, where the share of geotagged tweets is still close to one percent of all messages. For various purposes, it will be very useful to provide a method to geolocate as many messages as possible (Scheffler 2014; Fuchs et al. 2013).

To extract geolocation information from text in twitter messages, many different solutions have been developed. It is possible to get the location with a gazetteer. This could be difficult because of spelling errors, abbreviations etc. and even make matching impossible at times (Zhang and Gelernter 2014). To avoid this problem, neural networks are trained with a big dataset of messages, which are able to identify the text based location. This technique is known as named entity recognition (NER). Further on different string matching algorithms could provide a solution (Gelernter and Balaji 2013). Moreover the user, his personal social connections and information like the time zone or a specific usage of words are usable for the location prediction (Mahmud et al. 2012; Gontrum and Scheffler 2015).

However, most of these methods have things in common, that make them hard to apply in certain scenarios. Firstly, they need a very big dataset to evaluate the location and to train the neural network, e.g. all tweets in English language, and secondly, the location prediction is on a low level of detail. Our goal is to develop a method to extract the location for city districts, streets and points of interest in the city of Rostock, which means a rather small dataset is used and the level of detail which is needed is high. We address this problem in two different ways. First, we decided to choose an event, the Hanse Sail, with around one million visitors (NDR 2016). For a city like Rostock with around 210,000 inhabitants (Hansestadt Rostock 2015), this widely extends the potential database. Second, we are using a self-designed gazetteer, which holds all common places in the city of Rostock. To extract the location information from the tweets, we developed a two-tier matching method with the help of fuzzy text matching. Further, our target is to analyze the content as well as the sentiment of the extracted tweets.

3 Geolocation Identification

The algorithm to get the geolocation from each tweet is the core component of our work (see Fig. 1). First, we have to collect freely available tweets with the twitter streaming API. Second, we need to revise and normalize each message. We also want to identify the sentiment. For this, an extraction of laughter, smileys and emoticons is needed. The third part is the location itself. The platform which was used for all database operations is an Ubuntu powered server with 8 GB of RAM and an INTEL Xeon E5 2640 CPU with 2.5 GHz.

3.1 Collecting Tweets

To collect the tweets, we set up a harvester with the help of the python library Twython as well as the psycopg2 library for database connections (Mcgrath 2016; Gregorio and Varrazzo 2015). The harvester is connected to the twitter streaming API. The harvesting method is roughly based on Scheffler (o.J.). To get only German tweets, the tweets are filtered by a list with the most common German stop-words[1] and the twitter intern language identification. With this strategy, it is possible to harvest nearly all German language tweets in the timeframe of the Hanse Sail 2016 (10.08.2016–16.08.2016). Because of the relative small amount of German tweets, we are not running into the rate limit of the twitter gardenhose API. Each tweet is returned as a JSON-file. These files are parsed and stored in a

---

[1]http://www.ling.uni-potsdam.de/~scheffler/twitter/twython-german-stopwords.txt.

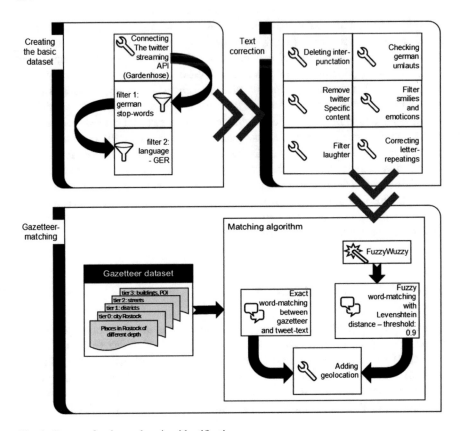

Fig. 1 Strategy for the geolocation identification

Table 1 Parsed JSON elements for the analysis

JSON element	Reason for parsing	Datatype
Text	Text of the twitter message	String
Date	Timestamp to extract posting time of the message	Timestamp
User id	To get user information of extracted tweets	Big integer
User name	To get user information of extracted tweets	String
Retweet	To filter for retweeted messages	Boolean
Latitude	Coordinates for each message	Double
Longitude	Coordinates for each message	Double
Place	Place information of the tweet	String

PostgreSQL database (see Table 1), which is enhanced with the PostGIS extension (Postgis Project Steering Committee 2015) being essential for the integration of a gazetteer.

3.2   Language Correction and Normalization

An important component in the process of analyzing social media messages is the text revision and normalization in order to compensate typing errors, smileys and written laughter. To address these issues, we developed a revision module.

3.2.1   Finding Smilies, Emoticons and Laughter

Smileys and emoticons are a very important part for the sentiment analysis later on. To recognize each smiley, the algorithm has to identify specific sequences of brackets, parenthesis and letters. For this the algorithm developed by Ritter et al. (2011) is very useful. For the emoticons, it is necessary to identify the unicode symbols and assign them a specific mood.

For the identification of written laughter in German language we assumed, that the first letter have to be an 'h'. Then the word has to be checked for the number of different characters. If there are only two different characters, 'h' has to appear at least as often as the other one. With this, words like 'haha' or 'hehe' are identified. Are there more than two different characters, 'h' has to be the most frequent character, so words like 'hihaheho' are also recognized. The recognized word is replaced with '%LAUGHTER'.

3.2.2   Correcting Letter Repetitions

Repetitions of characters, like 'Haaaaaalloooo' are very common in twitter messages. However these words possibly contain valuable information. To process these, the python library Natural Language Toolkit (NLTK) was imported (Bird et al. 2009). This library offers the opportunity to cut all character repetitions down to three. Then, for each word the repetitions are identified and all possible variations of the word are checked against a German dictionary with PyEnchant and, if one of the variations was found, replaced (Kelly 2014).

3.2.3   Dealing with Twitter Specific Content

There are several elements, which are specific to twitter messages only. Email addresses, links and retweet-signs are deleted completely. Messages which are addressed to users or topics are highlighted by the @-character. Therefor the @-character was removed but the word was preserved, because it often contains useful information for location purposes. The same procedure was applied to words which are starting with the #-character (hashtag).

3.3   Setting up the Gazetteer

The gazetteer is the core component of the matching algorithm. The gazetteer mainly integrates two databases: OpenStreetMap (OSM) and district borders from opendata.HRO. Specific Hanse Sail sites are manually integrated, like popular ships or the entertainment stages. Each gazetteer element was checked regarding its uniqueness inside Germany and inside Rostock. For the matching algorithm later on it is also important to know, if a LIW could have another meaning or a fuzzy check could provide many false positives. For example 'Evershagen' has to match exactly the word in the text, otherwise words like 'Leberversagen' (liver failure) or 'Herzversagen' (heart failure) are assigned to this district in Rostock using fuzzy checks. To know how accurate a location is, four depth levels are introduced (c.f. Fig. 1). The first depth tier describes the city scale. The second tier is for districts, the third for streets and the fourth for buildings, points of interest etc.

3.4   Gazetteer Matching

First, all geocoded tweets within a bounding box are sampled. Second, all German-wide unique gazetteer elements are compared with the text of each tweet. If a matching with a higher depth level was found, the other one is replaced. Because of the unstructured twitter language, the python library FuzzyWuzzy as well as python-Levensthein was integrated into the script (Cohen 2016; Haapala 2014). With those two libraries, a fuzzy string matching is possible to get as many locatable elements as possible. Through the fuzzy matching the element with the highest depth level is also chosen.

The next step is to get the non-unique elements as well. To achieve this, the result of the first matching is stored in a result table. All tweets in this result table are now compared with fuzzy matching and the non-unique elements from the gazetteer. If an element with higher depth level was found, the old location is replaced by the more detailed one. Some entries have to be verified and should be deleted if they do not match. So the last step is to check exceptions for some entries. For example, 'Alter Strom' is very often associated with electricity, not with the harbor district in Rostock Warnemünde. To delete these entries, special filters are created. In this case the tweet is checked against other words in the electricity context and if they match, the element is removed from the result table.

4   Results

From the 6.2 Mio. German tweets in the timeframe from 10.08.2016 to 16.08.2016 we were able to filter about 3,978 tweets which are related to the city of Rostock or a place inside the city. From those messages, 941 contain the words 'sail' or 'hanse', so it can be assumed that these are related to the Hanse Sail 2016. But there are also other relevant topics (see Fig. 2). It is interesting, that football/soccer is one of the most common topics, 13.5% of the messages are related to the FC Hansa Rostock. Another interesting outcome is, that tweets about holidays are not very common at that time (1%). Further, it shows, that the top five users do not provide the majority of the tweets. Only for the football/soccer topic 36% of the messages are provided by the top five users. The reason for this are live ticker which tweet throughout a match. In total, 500 different users were tweeting about the Hanse Sail. It can be concluded, that the distribution is not heavily influenced by automated and institutional tweets (Fig. 3).

To visualize the distribution of the tweets, all locatable tweets of depth two and three which are related to the Hanse Sail are drawn on a map with the help of a point density algorithm. From the 941 messages 91 are locatable on depth level one, 59 on level two and 57 on level three. These depth levels are chosen because the localization at lower depth levels is not able to show the hot spots inside the districts. Thus, 116 messages were used in the input dataset of the point density map. It shows that the most important place for the Hanse Sail 2016 is the city port of Rostock. But there are also hotspots in Warnemünde and near the international port. It is also noticeable, that well known ships are locatable, like the Krusenshtern or the Tolkien.

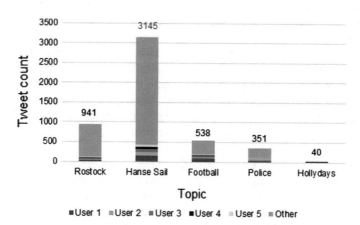

Fig. 2   Count of tweets for different topics with the most common users

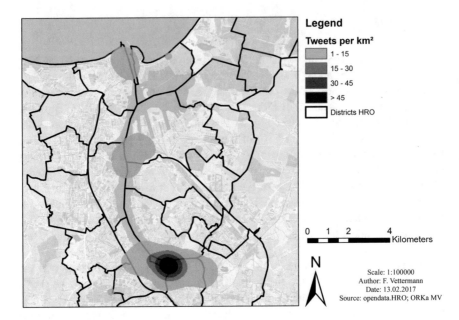

Fig. 3 Point density map of the tweets of depth level two and three in the city of Rostock

5 Discussion

We are able to extract tweets from the Hanse Sail 2016 in Rostock and to localize a part of them. This means there are still problems. First, we are only able to locate 3% of all the tweets at medium and high precision. If we include the tweets at the city district level, there are 207 messages with location information. Through manual checks we found, that not each locatable tweet was located by our algorithm. The problem is, that we had to enhance the gazetteer with more words. A good example for this is provided by Zhang and Gelernter (2014). They integrated toponyms from many more different databases. Colloquial expressions or abbreviations for places, like "Dobi" for "Doberaner Platz" as well as historical names, which could be very useful for the gazetteer, are also important (Walter and Bill 2015).

Li (2016) mentioned, that Volunteered Geographic Information (VGI) is a possibility to involve the younger generation into urban planning. With our data, we are able to provide valuable data source for decision makers. But this information can only be an extension for other data sources in urban planning. It was only possible to approve that the most important spot of the Hanse Sail is the city harbor Rostock. Only with a much bigger dataset, for example over three years, we are able to provide more reliable information about other hot spots through the Sail, like the ferry terminal.

Moreover, there are still problems with typing errors, abbreviations and the unstructured language in general. Other authors decided, to integrate abbreviations into their classification system as well and using neural networks for better results (Zhang and Gelernter 2014). The integration of abbreviations could also be very useful for us, but neural networks need training datasets to work properly. This could be a big problem in a region where many locations never have been mentioned in messages (Steiger et al. 2015). Another approach is to use other fuzzy algorithms to match the toponyms of the text with the gazetteer. Here it should be possible to use other functions, like the Jaro-Winkler-Distance which is available in the pg-similarity library, to compare it to the actual results (Oliviera 2012).

Finally, we have to think about performance issues. For the database of more than six million tweets used in this study, the algorithm needs almost 72 h to extract all locations. A more powerful platform or a smarter strategy matching the gazetteer could be essential for a planned live integration for the localization of messages.

6 Summary and Outlook

The developed algorithm is able to locate tweets in a short time frame with a high precision in the city of Rostock. With this tool, we demonstrated the possibilities which are provided by a social media analysis. The algorithm is able to extend the database of locatable tweets in Rostock massively. We are also able to show that the algorithm is able to address specific topics. This is very important especially for urban planning. For the Hanse Sail 2016 hotspots of the visitors are visible. From this data, information about interesting and important places for the visitors can be extracted.

We plan to extend our tool with other social media sources like Instagram to enhance our database. Further, we try to transfer our algorithm to a live analysis. Through this, we want to provide a decision support tool for urban planning purposes, which will also include a feature sentiment analysis module. The text analysis module will be enhanced to build relations between locations and topics respectively moods.

References

Agarwal A, Xie B, Vovsha I, Rambow O, Passonneau R (2011) Sentiment analysis of twitter data. In: Proceedings of the workshop on language in social media, pp 30–38

Bird S, Klein E, Loper E (2009) Natural language processing with python. O'Reilly, Sebastopol

Cohen A (2016) Fuzzywuzzy

Frias-Martinez V, Soto V, Hohwald H, Frias-Martinez E (2012) Characterizing urban landscapes using geolocated tweets, pp 239–248

Fuchs G, Andrienko N, Andrienko G, Bothe S, Stange H (2013) Tracing the German centennial flood in the stream of tweets, pp 31–38

Gelernter J, Balaji S (2013) An algorithm for local geoparsing of microtext. GeoInformatica 17(4):635–667

Gontrum J, Scheffler T (2015) Text-based Geolocation of German Tweets. In: Beißwenger M, Zesch T (Hrsg) Proceedings of the NLP4CMC 2015 Workshop at GSCL, Selbstverlegung, Duisburg, pp 28–32

Goodchild MF (2007) Citizens as sensors: The world of volunteered geography. GeoJournal 69(4):211–221

Grahem M, Hale SA, Gaffney D (2013) Where in the world are you? Geolocation and language identification in twitter—professional geographer 2013

Gregorio FD, Varrazzo D (2015) psycopg2

Haapala A (2014) python-Levenshtein

Hansestadt Rostock (2015) Statistisches Jahrbuch Hansestadt Rostock 2015, Selbstverlegung

Kelly R (2014) PyEnchant

Li C (2016) Social media: an ideal tool for public participation to promote deliberative democracy: the case of public participation in refugee crisis. Int J J Commun 1(2):36–41

Li R, Lei KH, Khadiwala R, Chang KC-C (2012) TEDAS: a twitter-based event detection and analysis system, pp 1273–1276

Mahmud J, Nichols J, Drews C (2012) Home location identifiaction of twitter users

Mcgrath R (2016) Twython

NDR (2016) Hanse sail: positive bilanz trotz regenwetters. http://www.ndr.de/unterhaltung/events/Hanse-Sail-Positive-Bilanz-trotz-Regenwetters,hansesail896.html (Stand: 2016-08-15) (Zugriff: 2017-02-07)

Oliviera ETD (2012) PG-similarity

Postgis Project Steering Committee (2015) PostGIS: PostGIS project steering committee

Ritter A, Clark S, Mausam, Etzioni O (2011) Named entity recognition in tweets: an experimental study. In: Proceedings of the 2011 conference on empirical methods in natural language processing, pp 1524–1534

Scheffler T (o.J.) Linguistics of German twitter. http://www.ling.uni-potsdam.de/~scheffler/twitter/#corpus (Stand: 2016-06-01) (Zugriff: 2016-06-30)

Scheffler T (2014) A German twitter snapshot. In: Calzolari N, Choukri K, Declerck T, Loftsson H, Maegaard B, Mariani J, Moreno A, Odijk J, Piperidis S (Hrsg) LREC 2014, ninth international conference on language resources and evaluation: May 26–31, 2014, Reykjavik, Iceland, proceedings. [S.l.]. European Language Resources Association (ELRA)

Steiger E, Resch B, Zipf A (2015) Exploration of spatiotemporal and semantic clusters of Twitter data using unsupervised neural networks. Int J Geogr Inf Sci 1–23

Traynor D, Curran K (2013) Location-based social networks. In: Lee I (Hrsg) Mobile services industries, technologies, and applications in the global economy. Information Science Reference, Hershey, pp 243–253

Walter K, Bill R (2015) Aufbau eines crowdsourcing basierten verzeichnisses für historische ortsnamen. In Strobl J (Hrsg) AGIT-symposium 27 (Salzburg)// geospatial minds for society 2015. AGIT—journal für angewandte geoinformatik 1-2015. Herbert Wichmann Verlag; Wichmann, Berlin/Offenbach, pp 432–439

Zhang W, Gelernter J (2014) Geocoding location expressions in twitter messages: a preference learning method. J Spat Inf Sci 9

Smart Monitoring System of Air Quality and Wall Humidity Accompanying an Energy Efficient Renovation Process of Apartment Buildings

Grit Behrens, Johannes Weicht, Klaus Schlender, Florian Fehring,
Rouven Dreimann, Michael Meese, Frank Hamelmann,
Christoph Thiel, Thorsten Försterling and Marc Wübbenhorst

Abstract In this joint project the development of a robust low-cost monitoring system of air quality and wall humidity values is presented. Using open source software on a RaspberryPi 3 with some plug-in sensors for temperature, pressure, CO_2 and humidity data sets every minute are stored in an universal data base. Data analysis gives good results in recognition of ventilation states inside the apartment, which can be used to give ventilation advices to the apartment users in time of renovation, to avoid bad air quality and mould groth. For protection of personal room user data a NTRU Public Key Cryptosystem is developed and implemented.

1 Introduction

The joint project of a group of researchers from FSP IFE (research focus "Interdisziplinäre Forschung für dezentrale, nachhaltige und sichere Energiekonzepte") from Bielefeld University of Applied Sciences and the renovation management in Bielefeld-Sennestadt developed a smart monitoring system for air quality and wall humidity accompanying the renovation process in outdated apartment houses. The houses with relatively small flats are built in the fifties and sixties especially for refugees from eastern European countries and they are occupied mostly by people with relatively low-wage stage, for example pensioners or single mothers with children. Field studies including data monitoring in comparable building stock were provided by (Cali 2016), where the hardware of the

G. Behrens (✉) · J. Weicht · K. Schlender · F. Fehring · R. Dreimann · M. Meese ·
F. Hamelmann · C. Thiel
FSP IFE—University of Applied Sciences Bielefeld, Artilleriestraße 9, 32427 Minden,
Germany
e-mail: gbehrens@fh-bielefeld.de

T. Försterling · M. Wübbenhorst
alberts. architekten BDA, Marderweg 21, 33689 Bielefeld, Germany

© Springer International Publishing AG 2018
B. Otjacques et al. (eds.), *From Science to Society*, Progress in IS,
DOI 10.1007/978-3-319-65687-8_16

monitoring system was very high level and cost expensive. The global aims of the here presented prototype of smart monitoring system are the ability of application in larger amounts, personal data protection and an intelligent analysis of the recorded data. The data analysis results of the here described two-years-project will be the point of beginning of research the user behavior with focus on "Rebound effects after renovation processes". Three flats in multi-apartment houses and one detached house were monitored over one year time. The project is supported by ministry of North Rhine Westphalia (2016–2020) and by the KfW program "Energy-efficient Urban Redevelopment" (number 432) (BMUB 2016).

2   Measuring Data and Sensor System Technology

A standalone measuring system based on RaspberryPi with Debian Linux is used to monitor the different appartments. The sensors are connected by I2C databus (Cali 2016). For specific observations of the different environmental conditions, several sensors are needed. For the measurement of the relativity humidity RH_{in}, pressure p and air temperature T_{air} is recorded with a Bosch Sensortec Integrated Environmental Sensor Unit "BME280" two 1-Wire digital thermometer "DS18B20" for the inside temperature of the outer wall (T_{wall}) and of the heating (T_{heat}). The position of temperature sensor (outer wall) is placed, where the coolest temperature are expected. Carbon dioxide levels (CO_2) are recordet by a Telaire T6613 single channel CO_2 Module. Based on air pressure, air temperature and relativity humidity, the vapour pressure and therefore the absolute humidity can be calculated as shown in (Sonntag 1990). With absolute and relative humidity of the inside wall, the relative humidity of the outer wall can be approximated backwards using wall temperature as shown in Fig. 1.

Fig. 1 Sensor locations for an example one-room-apartment. The apartment consists one temperature sensor for heating, one temperature sensor for the outside wall temperature, CO_2. Air pressure, humidity and temperature sensors are in the middle of the apartment

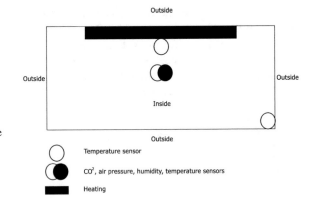

3   Data Monitoring System

As seen in Fig. 2 the measuring system in apartments consists of many sensor computers (RaspberryPi's). Each of them collects measurement data of sensors and actuators and sends them by POST request to a RESTful Webservice interface, which the main server controls (Fig. 2 left side). The main server handles any type of data and saves it into a local database. The amount of sensor data and standalone computers is scalable. It is possible to have an unlimited amount of measuring systems which send encrypted data from each main server of an apartment to our FH-Server (Fig. 2 right side). We have a 1:n relationship between a data receiving server and a scalable amount of data serving sensor computers in an apartment. Also we could have a scalable amount of data collecting apartements which send NTRU encrypted data to the FH-Server by calling a special webservice.

The FH-Servers host a web application for data collection, analysis and visualization with an architecture as shown in Fig. 3. The frontend is based on HTML5, CSS, JQuery and JavaScript with client side communication by JSON server interfaces. In focus of backend application is an efficient communication and storage of all data types. The implementation of algorithms for large data sets and data analysis allows us to support big data amounts in the system. In Backend Application standard open source technologies of Java EE, such as EJB, JNDI, JPA, JAXB, JAX-RS are used with RESTful webservice interfaces to manage data communication. In the JSON based client—server communication via Web service. For high security level a NTRU kryptography support was implemented.

A visual display unit as seen in Fig. 4 is located inside monitored appartment rooms. Temperature, relative huminity and carbon dioxide level (CO_2) are visualized. Also a feedback by room users of their feelings about air quality is requested. Feedback-values of green, yellow and red coloured smiley-icons can be used for supervised training of machine learning algorithms in the future.

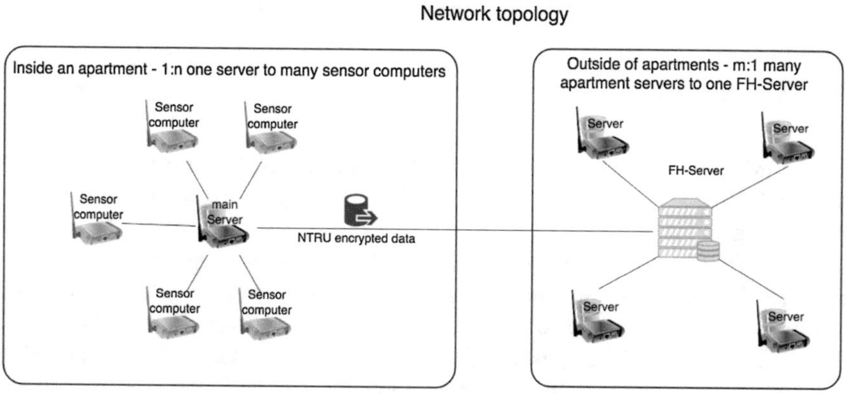

Network topology

Fig. 2 Network topology of sensor systems

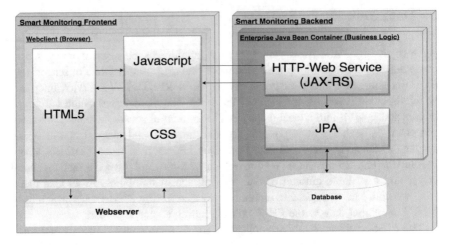

Fig. 3 Web application architecture

Fig. 4 Visual display unit with visualization of air quality measurement and Feedback buttons

4   IT Security

To ensure maximum privacy all data packages sent via Internet are encrypted using the NTRU Public Key Cryptosystem (Hoffstein 1998), which is implemented and parameterized for this case.

The NTRU Cryptosystem is a lattice based, asymmetric cryptosystem, in which the security is based on finding the shortest, non trivial vector in a lattice (SVP) (van Emde Boas 1981; Ajtai 1996). SVP is proven to be NP-hard for randomized lattice reductions (Ajtai 1996). After Google's announcement to build a large quantum computer, in the near future one can feel the need of using a cryptosystem, which is secure against attacks from a quantum computer. Typically, the NTRU Public Key Cryptosystem is implemented as a hybrid cryptosystem that means that you also need a symmetric cryptosystem for encoding data. One encrypts the data symmetrically with a session key and encrypt the session key with the public key of the asymmetric NTRU Public Key Cryptosystem. This is a lot faster if the data packages are quite large. We use the NTRU Public Key Cryptosystem in a different way. Because only small data packages exists, a symmetric cryptosystem like AES is chosen, which normally speeds up the encrypting process.

As defined three sensors and three different data packages have to be encrypted. Relative humidity, pressure and temperature are recorded as double values by the BME280, the CO_2 level is recorded by the Telaire T6613 as an integer. Wall temperature and heating temperature are recorded as an double value. A timestamp as long and an id as integer value is added to the data.

Recorded data were sent through a post quantum save tunnel over the internet without an symmetric encryption. The possible decryption failure, which can happen as published by (Hoffstein 1998), could be successful eliminated.

The implementation of NTRU algorithm and RESTful Webservices for data-transfer is able to handle about 50 datasets per minute on a RaspberryPi 3. One apartment delivers approximatly 1,555,200 datasets per year, see Fig. 5. With NTRU based data transfer 3000 datasets per hour can be transmitted. This is only 5,7% of possible NTRU live data transfer to the FH-Server under standard data transfer rate via Internet. To scale up the amount of persisted and transferred datasets by our monitoring system up to 105 sensors of each apartment is possible.

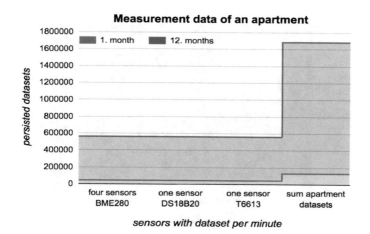

Fig. 5 Amount of data

5 Data Analysis

The CO_2-value is used to determine the quality of the air inside of the observed apartments. The lowest CO_2 level is expected to be 450ppm, as found outside of the observed building. The threshold, when ventilation should be initiated, is 1000ppm above the outside level based on (Emde Boas 1981).

To give a suggestion for improving the manual ventilation to the room user, the ventilation suggestion value SCO_2 is introduced as:

$$s_{CO2} = \frac{CO_2 - 450\text{ppm}}{1000\text{ppm}}$$

The room users handled the ventilation manually. Active air circulation is detected by using the measured values of CO_2 level. If the slope of the CO_2 level is higher than zero, no air circulation is active. This is defined as the status "window closed". If the slope value of CO_2 is smaller than zero, an air circulation is active— status "window open". During our measurement at one-minute intervals slope values smaller than zero were also observed, when no window was open. More exactly these values were between 0ppm/min and −5000 ppm/min. These values are made by "infiltration", e.g. small lacks due to the age of the mounted frames of the windows. All values lower than −5000ppm/min are defined as "window open". The slope is calculated by neighboring data sets in time series with measuring intervals of one minute. The graphical result of this analysis is shown in Fig. 6.

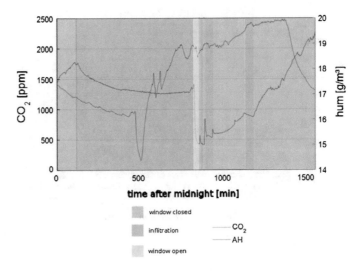

Fig. 6 Absolute humidity and CO_2 level of a random typical day. The CO_2 level is most of the time under suggestet level of about 1500ppm

Fig. 7 Air quality factor "s" of CO_2 level and humidity: If one parameter exceeds one or bigger (*darker areas left* and *right side*), ventilation should be initiated. If both paramters are lower than one (*lighter middle area*), no ventilation is needed

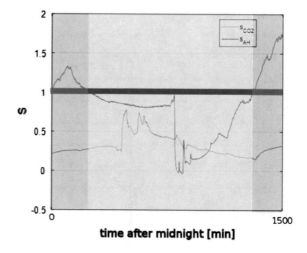

time after midnight [min]

The proposed ventilation (bayed on humidity) "S_H" can be approximated as:

$$s_H = \begin{cases} \left| \frac{AH_{in} - 10.1g/m^3}{AH_{out} - 10.1g/m^3} \right| & \text{if } AH_{out} \neq \frac{10.1g}{m^3} \\ 1 & \text{if } AH_{out} = \frac{10.1g}{m^3}, \end{cases}$$

where **AH** is the absolute air humidity. RH_{in} should be 50% at 22 °C, or 10.1 g/m^3 as the absolute humidity **AH** should be used.

This configuration was tested for several months during summer time (May 2016–August 2016) and winter time (December 2016–March 2017). Each dataset was stored in smart monitoring system. Results for the calculated s_H can be seen in Fig. 7.

Combining the parameter s_{CO_2} and s_H can be concluded as an index for the overall air quality **S**:

$$S = \max(s_{co_2}, s_h)$$

If the value of S is lower than one, the air quality is acceptable and no ventilation needs to be activated by e.g. opening a window. If one of the observed parameters is higher than 1, the ventilation should be switched on. This result is shown in Fig. 7.

6 Humidity Outer Wall

The inner surface of the outer wall is an important parameter. If the outer wall is not isolated enough, energy loss can happen in combination with mold-infested surfaces. Besides discomfiture, this infestation can cause health issues. Therefore, the

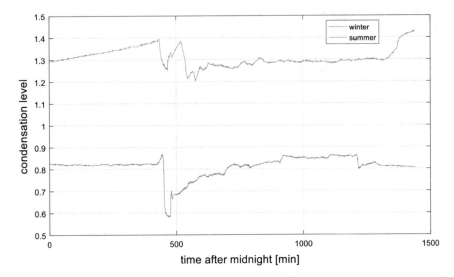

Fig. 8 Relative humidity of outside wall during a random winter (*upper line*) and a random summer day (*lower line*)

surface temperature is an important factor to determine the surface humidity. The relative air humidity, pressure and air temperature are measured. By these parameters the absolute humidity can be calculated. Assumed, that this absolute humidity is the same in every point of the room, the relative humidity of the inner surface of the outer wall or the condensation of this surface based on (Sonntag 1990) can be determined. If the condensation is higher than 1, the surface gets wet. As shown in Fig. 8, the condensation is higher than one in winter time. During this period, the condensation wetted the surface and it can get mold-infested. In summer time condensation is lower than one. Therefore, the water does not condensate at the inner surface.

7 Results and Discussion

A robust and low-cost sensor network system based on RaspberryPi platform was developed for measuring air quality and wall humidity in appartment houses. The monitoring software system has been designed and developed by using Java Enterprise Edition web technologies. The number of sensor data from sensors and their applications could be stored as efficiently as possible by RESTful Webservice interfaces. Data transfer of personalised data via Internet is highly secured by implemented post quantum safe NTRU crypto algorithm. The data is processed by algorithms and is stored into an efficient universal database. The developed data measuring system on Backend offers scalability, maintainability and extensibility to handle a very big amount of sensors. An easy exchange, add or even expand of system parts is ensured. It provides a solid base to manage long-term monitoring data, which is prepared now for more expermients in larger amounts. An easy to use display unit for vizualization of current air quality values and for getting user feeling feedbacks is prepared. It ensures a feedback for further automatic regultaed air ventilation and heating systems monitored by our measuring system.

Results of data analysis before the renovation process shows, that the developed low-cost smart monitoring system for outdated appartment houses is able to recognise ventilation states. It is ready to accompany a renovation process with monitoring the user behaviour and giving advises in ventilation and heating to room users to avoid mould growth after renovation just in time.

The smart monitoring system is also able to give commands to an automatic ventilation and heating system during and after energetic renovation of the apart-ment houses. In future works a new academical research project based on the developed smart monitoring system in Bielefeld-Sennestadt for active assistance of the users during renovation process is in final preparation.

References

Ajtai M (1996) Generating hard instances of lattice problems. In: Proceedings of the twenty-eighth annual ACM symposium on Theory of computing, Philadelphia, 22–24 May 1996

BMUB (2016) Energetische Stadtsanierung. http://www.energetische-stadtsanierung.info. Accessed 4 Apr 2017

Cali D et al (2016) Beschreibung des feldversuchs. In: Cali D, Heesen F, Osterhage T, Streblow R, Madlener R, Müller D. (eds) Energieeinsparpotential sanierter Wohngebäude unter Berücksichtigung realer Nutzungsbedingungen. EnEff: Stadt, pp 38–42. Fraunhofer IRB, Stuttgart

DIN (2007). DIN EN 13779: 2007–09 Lüftung von Nichtwohngebäuden–Allgemeine Grundlagen und Anforderungen an Lüftungs-und Klimaanlagen und Raumkühlsysteme. Berlin

Hoffstein J et al (1998) NTRU: a ring-based public key cryptosystem. In: International algorithmic number theory symposium. Springer, Berlin Heidelberg, pp 267–288 (1998)

NXP Semiconducters: UM10204 I2C-bus specification and user manual (2017). http://ww.nxp.com/documents/user_manual/UM10204.pdf. Accessed 1 Apr 2017

Sonntag D (1990) Important new values of the physical constants of 1986, vapour pressure formulations based on the IST-90, and psychrometer formulae. Z Meteorol 70(5): 340–344

van Emde Boas P (1981) Another NP-complete partition problem and the complexity of computing short vectors in a lattice. Technical report 81-04 https://staff.fnwi.uva.nl/p.vanemdeboas/vectors/mi8104c.html. Accessed 1 Feb 2017

A Lightweight Web Components Framework for Accessing Generic Data Services in Environmental Information Systems

Eric Braun, Alessa Radkohl, Christian Schmitt, Thorsten Schlachter
and Clemens Düpmeier

Abstract This article presents a lightweight framework that allows web components to access data services via standard interfaces. The framework is used in combination with the Generic Microservice Backend (GMB) project which provides a wide range of different services for data management, analysis and visualization e.g. for the implementation of web based environmental information systems. The framework allows an easy implementation of web components by using a centralized communication and controller module that can handle interactive events in the frontend of a web application and data requests towards the backend. The framework can be used to create complex dashboard and portal server pages with a responsive design without much programming.

Keywords Web components · Environmental information systems · Data services · Generic microservice backend (GMB)

1 Introduction

The Web based Information System Team (WebIS) of the Institute for Applied Computer Science (IAI) at the Karlsruhe Institute of Technology (KIT) performs research on the usage of innovative Internet technologies for implementing large scale information systems for application fields, such as environmental information systems, collaborative scientific data and information platforms, smart energy system solutions or Internet of Things (IoT) applications.

In the last years, the focus lied on conceptualizing and implementing an innovative and very generic data management infrastructure, which can be reused in many application contexts. This microservice architecture based framework called the Generic Microservice Backend (GMB) uses Docker containers as runtime

E. Braun (✉) · A. Radkohl · C. Schmitt · T. Schlachter · C. Düpmeier
Institute for Applied Computer Science, Karlsruhe Institute
of Technology (KIT), Karlsruhe, Germany
e-mail: eric.braun2@kit.edu

© Springer International Publishing AG 2018
B. Otjacques et al. (eds.), *From Science to Society*, Progress in IS,
DOI 10.1007/978-3-319-65687-8_17

environment for generic data management services, which can be independently scaled and distributed onto different runtime infrastructures, such as Cloud and/or Big Data infrastructures, private computing clusters with container runtime support, e.g. based on the OpenStack runtime environment, or even single server or development laptops.

GMB services are already used in several environmental information systems, such as the environmental information portals of several German states based on the LUPO portal framework, e.g. the environmental information portal of Baden-Württemberg[1] where the GMB is part of the server side Web Cache for providing fast access to environmental data of different types (e.g. for serving geo data, measurement data, text and binary assets and different forms of master data) to environmental information portals but also to mobile environmental applications. Furthermore, the GMB is used for other applications, like energy system applications as required for the Energy Lab 2.0 and Energy System 2050 (Düpmeier et al. 2016; Hagenmeyer et al. 2016) projects.

While the functionality provided by the GMB services can be easily accessed from client applications by using well-defined REST-based service APIs, in the past, each application using the GMB developed its own dedicated client side data access, application and display logic for displaying GMB data or accessing data driven services. An analysis of the corresponding code in different applications revealed that this code is largely redundant across different applications. Even the presentation logic is often similar and visualizations of the data or interactive data forms for gathering data only differ in minor aspects and style. Therefore, the question arose if application development benefits from a client framework that unifies the code for accessing, processing and visualizing service provided data in client side applications.

Since applications using the GMB are mainly web based or hybrid apps, in a first step, different web technologies were evaluated to find a suitable approach and base technologies for setting up a JavaScript and HTML based web client application framework for handling GMB data in client applications. Since different applications using the GMB services use different JavaScript frameworks, e.g. AngularJS or Reactive.js, for implementing the web application logic, the evaluation quickly showed that the focus lied on standardizing on one of these frameworks or even building a new generic JavaScript framework for web application development. Instead, it can be stated that all modern JavaScript web application frameworks going the direction of using modern web component technology for allowing application developers to build their own complex UI components which can then be reused for different applications. Therefore, the basic idea was to conceptualize and develop a lightweight web framework which can provide an easily extendable library of reusable web components for accessing and presenting GMB data which can be used in different web application contexts side by side with frameworks like

---

[1]https://www.umwelt-bw.de/.

AngularJS and Reactive.js, or in more simple use cases by just using only the developed framework and e.g. jQuery.

This article describes the WebIS Web Components (W2C) framework in more detail and how it interacts with the GMB service infrastructure. Additionally, some examples of components and applications are shown that demonstrate the usability of the components for a broad range of environmental applications. Finally, a conclusion is drawn and an outlook completes this article.

2 Generic Microservice Backend (GMB)

In this chapter only a brief overview on the GMB is given. More details on the GMB can be found in Braun et al. (2017) and Schlachter et al. (2017). Figure 1 shows how the Generic Microservice Backend (GMB) is divided into several distinct services which can be accessed by an integrated REST-based service API provided by a gateway. The web component framework (W2C) can be seen as JavaScript client framework for accessing the GMB in applications through a server side gateway.

The GMB provides a bundle of separated and scalable microservices for data management, data analysis and visualization that are access controlled, load balanced and integrated through a central gateway as external access point. Figure 1 shows that each data management service uses its own database to store different types of data in e.g. different types of database management systems (polyglot data management). The *Master Data Service* manages structured objects that describe real objects, such as a nature protection area, lake or river, or different assets of a company, e.g. a measurement station, measurement equipment or a sensor. Typically, this kind of data can be stored in a NoSQL document oriented database. The *Time Series Service* manages time series data using dedicated databases that are optimized for typical time series queries and operations. The *Digital Asset Service* is responsible for files that can be stored as separate documents together with associated metadata. E.g., the persistence layer of this service can consist of one or more content or asset management systems, such as Alfresco.[2]

Beside these data services, Fig. 1 shows two additional services which allow to store semantic information about the data contained in the data services. The *Schema Service* stores definitions of the common structure of master data objects (e.g. schema) and how structures interrelate. Schemas are stored in JSON schema format. The *Link Service* manages relationships between different data instances in the separate services. These can be relationships between different master data objects but also between a master data object and digital assets or time series. Together, these services allow the instrumentation of complex, integrated data

---

[2]https://www.alfresco.com/.

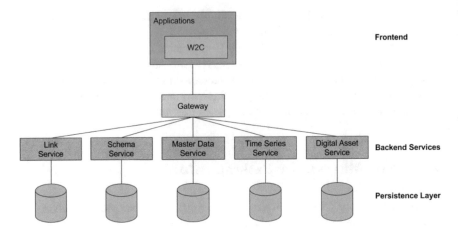

Fig. 1 Communication between the GMB services and the W2C through a central gateway

management models for applications on top of the GMB which cover a broad range of application types.

Each of these services has its own API (using REST Tilkov 2009), but all APIs are exposed through a common integrated API via the gateway. To allow standard access to the data services the GMB API, the gateway supports standard APIs such as Atom feed format or OpenSearch wherever possible. The W2C framework presented in the next chapter uses the gateway API to access the GMB services thereby also instrumenting standard APIs as much as possible. This allows compatibility of the W2C data interfaces with many other services providing the same or similar APIs.

3 A Lightweight Web Component Framework for Accessing Data Services

The basic architecture of the lightweight web component framework (W2C) for accessing GMB and other data services is depicted in Fig. 2.

One part is a library of reusable web components (W2C Components) which can be used by authors to integrate data presentations of GMB data into web pages without programming. The second part is the W2C Base which eases the development of (new) components by providing common client side background logic. The W2C Base has three major responsibilities. It decouples the data access to the services from the components by providing a client side data cache (data broker) as data access interface for components using an asynchronous interface for optimal performance. Additionally, it implements functionality to bootstrap and load Polymer and the basic W2C infrastructure for all components on a web page (W2C Base Class). Finally, it provides functionalities for coordination of several components by

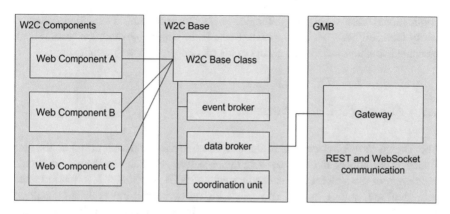

Fig. 2 Architecture of the W2C framework

providing event communication between loaded components (event broker) and exposing coordination functionality (coordination unit) via an API to coordination components. The last two features are necessary for coordinating several components on a web page so that they form a larger interactive single web page application.

With these functionalities in place, the W2C framework allows the aggregation of complex dashboard web pages that include several interactive components coordinated with each other by just assembling W2C web components on a page and configuring them for data access, coordination and event exchange. A click in one component can then e.g. trigger an update of data in the data broker and the arrival of the new data in the broker can result in a coordinated but asynchronous update of the information display of all web components having a dependency on these data.

In order to implement a new web component using W2C, the component just has to inherit from the W2C Base Class that exposes the basic functionalities to the component. The developer can then use all Polymer provided components, additional JavaScript libraries, the W2C provided functionality and already existing W2C components for implementation of their own component. The next chapter shows some sample components and how they are used in example applications.

4 Integration and Usage in Sample Applications

To gather requirements for the W2C concept several existing environmental information system applications were analyzed in advance to the development in relation on how web components enhance the usability of application developers and end users. After developing the basic concepts and first prototypical implementation these environmental applications were also used for the evaluation of the concept. For evaluation, parts of the existing information displayed in these applications were newly implemented with corresponding web components to verify the usefulness of the approach. In the following two evaluation scenarios are presented to show how W2C components can be used in web applications.

4.1 Theme Park Environment

The Theme Park Environment is the official environmental web portal of the state Baden-Württemberg in Germany with the objective of enhancing the environmental awareness of the general public Ruchter et al. (2003). It contains information about soil, geology, nature areas, and related environmental issues in the state of Baden-Württemberg in the southwest of Germany. It was developed by the Institute for Applied Computer Science of Forschungszentrum Karlsruhe (Karlsruhe Research Center) and the State Institute for the Environment, Measurements, and Nature Protection on behalf of the Ministry of the Environment of Baden-Württemberg and continuously enhanced since its launch in 2006 Düpmeier et al. (2006).

Since the web user interface of the Theme Park was not built for mobile access from the beginning and mobile usage is growing rapidly, a redesign of the Theme Park is foreseen with the aim to fully support mobile devices besides desktop browsers by implementing a responsive and adaptive web user interface based on the current HTML5 standard and aiming also for a better accessibility.

Therefore, the KIT explored in Radkohl (2016) if web component technology would be a good solution to fulfill these tasks by implementing responsive and adaptive web components for the main information display use cases. As information system for the general public the Theme Park instruments many different types of media presentations, such as images, panoramas, slideshows, streaming videos or interactive maps for information display. Thus, for the evaluation of the usability of web component technology for the Theme Park several media and web content display components were developed which could be used in the redesigned Theme Park to display binary media assets and web contents.

Figure 3 above shows an advanced adaptive and responsive component for displaying both web text content and images aligned on the right side of the text by using another component for the display of the images together with associated metadata. The combined component can display the images at different positions depending on provided configuration parameters. The author of the web page can choose between different layout templates to define how text and images are laid out together (the screenshot shows a "text left and image column right" layout). On smaller displays the component repositions the images accordingly to responsive design principles (see the right part of Fig. 3). In addition, the component allows displaying metadata like the title, author information and links that lead to more information about the displayed topic.

The following screenshot (Fig. 4) shows another web component which displays slideshows. This component uses the web component for image display mentioned

```
<text-component content-data="…"></text-component>
```

Fig. 3 Web component showing an article with accompanying images

```
<media-slideshow slideshow-data="…"></media-slideshow>
```

Fig. 4 Slideshow component which presents different images

above to embed this component into a slideshow viewer which cycles through the display of several media files integrated into one slideshow.

The slideshow component can be used in several contexts: for presentation of large scale header banners or smaller slideshows embedded into side columns.

Figure 5 shows a typical web page of the Theme Park as it is currently implemented. The mentioned web components can be used to re-implement key parts of this page by using the text-component for implementing the combined text and image part in the middle of the page in order to be more responsive and adaptable while the slideshow component can be used to provide a modern mobile friendly replacement for the slideshow component on the right side of the shown page which also understands gestures for scrolling through the images.

For the evaluation of the web components several pages of the Theme Park were recreated on a Liferay portal server by first creating a portlet wrapping the above-mentioned web components and the W2C framework, and then using the above discussed components for information display. The evaluation showed that all web pages of the current Theme Park could be easily recreated on a Liferay server by using only a few generic components for media and web content display. Combined with a new responsive corporate design based on e.g. Bootstrap a true responsive and adaptable web user interface can be created to implement the new version of the web user interface of the Theme Park, which would be fully responsible, adaptable and would feature even touch screen gestures for true mobile friendliness. All accessibility features can be implemented as inherent part of the web components so that the information displayed by the web components can also be used by screen readers and other browsers used by persons with disabilities.

Fig. 5 Typical theme park environment web page

4.2   Web Components for Displaying Measurement Data

Many environmental information systems including the Theme Park Environment have to display diagrams showing measurement or other time series data. Currently, often only images of such diagrams are embedded into web pages. However, this poses problems for responsive web applications, because these images of diagrams typically do not scale well for different device display sizes. Images also do not allow exploring measurement data in a more interactive way.

Therefore, as another use case it was explored if web components could be used to ease the handling of interactive responsive visualizations of time series data in web applications. This use case led to the development of the FlexVis framework Braun et al. (2016) which provides a web component and microservice-based framework for integrating time series data driven visualizations into web applications based on implemented visualization components and pre-configurations of component instances that define the mapping between the visualization components and background data services.

Figure 6 shows a data visualization of NO_2 pollution in the air being an example use case needed for visualizing air quality data in environmental data portals.

W2C is compatible with the visualization framework FlexVis Braun et al. (2016) and allows to use FlexVis components together with other W2C components for such visualizations. The FlexVis visualization components are directly configurable using the framework. Additionally, the data binding of the FlexVis components can instrument the data broker of the W2C framework, and FlexVis components can participate in W2C event communication.

One use case of such types of web components is the home page of the Landesamt für Umwelt, Messungen und Naturschutz (LUBW) which is responsible

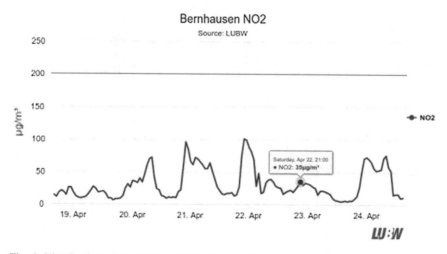

Fig. 6 Visualization of air pollution (NO_2) at a specific measuring station

for providing environmental measurement data for the German state Baden-Württemberg e.g. air quality. The LUBW plans to redesign their web pages e.g. providing information about air quality in a more responsive and interactive way. Early concepts for the new information display combine a map component for station selection, a selection component for physical properties and line charts for data visualization. Figure 7 on the next page shows an early prototype of the new web page. The map component, the line chart component and the property selection box are all implemented using W2C and use the data broker to fetch the relevant data for visualization.

Furthermore, the state of the different components is coordinated using the event broker and coordination unit. The user can select a measuring station on the map and the line chart shows the correct data for this station. Analogously, the property panel at the top can be used to change the physical measurement property the user is interested in which can result in a change of measurement stations shown on the map; and both the map component and the visualization component are updated correspondingly.

Both presented scenarios show, how the different features of W2C can be used. In general, W2C components can be mixed with other web contents to create portal- and dashboard-like web pages composed out of static web content coming from an e.g. portal or content management server and interactive components that can communicate with each other and backend services by using the appropriate brokers. Since web components are implemented to be responsive, adaptive and

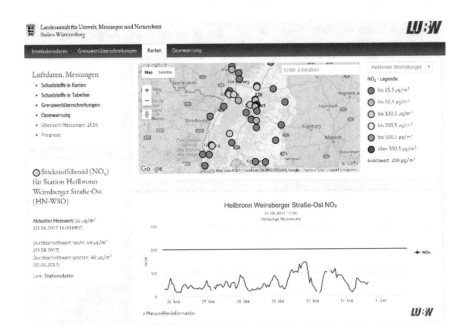

Fig. 7 New concept of LUBW webpage for air quality information

accessibility aware, their usage in web based applications simplifies the provision of more mobile friendly, fully accessible and highly interactive web pages.

5 Conclusion

The previous chapter showed that the W2C framework can ease the implementation of simple but also complex web applications. Furthermore, the article also showed, that the components can be used to instrument a large variety of use cases in different application scenarios. Using the event, data and coordination functionality of W2C, the web components can interact with the backend as well as with each other in a structured and controlled way. The Polymer library as well as the W2C base class help the developer to quickly create new components which provide the same elementary basic interaction elements. This harmonizes the interaction elements across different web components and makes the interaction possibilities more understandable for end users. Furthermore, many components exist that are ready to use and that already provide responsiveness, adaptability, accessibility, a unified design and high customizability.

In the future, W2C can provide even more web components for developers to choose from. Additionally, existing components can be enhanced for even more configurability. Finally, coordination configuration can be eased by implementing a special coordination configuration component.

References

Braun E et al (2016) Generic web framework for environmental data visualization. In: Advances and new trends in environmental informatics EnviroInfo 2016, Berlin, Germany, 14–16 September 2016, pp 289–299. doi:10.1007/978-3-319-44711-7_23

Braun E et al (2017) A generic microservice architecture for environmental data management. ISESS

Düpmeier C et al (2006) Theme park environment as an example of environmental information systems for the public. In: Environmental modelling and software 2006, pp 1528–1535. doi:10.1016/j.envsoft.2006.05.011

Düpmeier C et al (2016) A concept for the control, monitoring and visualization center in energy lab 2.0. In: Energy informatics. Lecture notes in computer science, vol 9424. Springer, pp 83–94

Hagenmeyer V et al (2016) Information and communication technology in energy lab 2.0: smart energies system simulation and control center with an open-street-map-based power flow simulation example. Energy Technol 4(1):145–162

Radkohl A (2016) Konzepte und Best Practices für die Nutzung intelligenter UI-Komponenten zur Darstellung von Inhalten in webbasierten Informationssystemen. Bachelor thesis

Ruchter M et al (2003) Web-based environmental information systems for the public: concepts, potentials, and applicability based on a case study

Schlachter T et al (2017) Generic web cache infrastructure for the provision of multifarious environmental data. ISESS

Tilkov S (2009) Rest and HTTP. dpunkt. verlag

From Sensors to Users—Using Microservices for the Handling of Measurement Data

Thorsten Schlachter, Eric Braun, Clemens Düpmeier, Hannes Müller
and Martin Scherrer

Abstract This article presents a microservice-oriented concept for the handling of measurement data. It consists of five phases providing collection, processing, persisting, dissemination and presentation of measurement data. The concept has been implemented and evaluated for one use case dealing with air pollutants and air quality time series. It includes tools for data ingestion as well as a set of generic microservices for the provision of different data types.

Keywords Data ingestion · Measurement data · Microservices · Sensors · Visualization

1 Introduction

Digitization is currently a hot topic in Germany which is often interpreted in the context of Industry 4.0 or Internet of Things (IoT) applications as gathering data via sensors, storing these data in e.g. big data infrastructures, analyzing data, aggregating further results and then using them for making smart decisions. In this political setting

T. Schlachter (✉) · E. Braun · C. Düpmeier
Institute for Applied Computer Science (IAI), Karlsruhe Institute of Technology (KIT),
Karlsruhe, Germany
e-mail: thorsten.schlachter@kit.edu

E. Braun
e-mail: eric.braun2@kit.edu

C. Düpmeier
e-mail: clemens.duepmeier@kit.edu

H. Müller · M. Scherrer
State Institute for Environment, Measurements and Nature Conservation
Baden-Wuerttemberg (LUBW), Karlsruhe, Germany
e-mail: hannes.mueller@lubw.bwl.de

M. Scherrer
e-mail: martin.scherrer@lubw.bwl.de

© Springer International Publishing AG 2018
B. Otjacques et al. (eds.), *From Science to Society*, Progress in IS,
DOI 10.1007/978-3-319-65687-8_18

the Environmental Agency of Baden Württemberg (LUBW) which is responsible for providing environmental information to third parties is exploring ways to modernize their handling for measurements in order to deliver data and analysis results timely to the end user via Web and mobile applications. The environmental sector is a complex interdepartmental and cross-sectional application field that has numerous relationships and/or interactions with other sectors. An example for the coupling of the environment with other areas is the supply with sustainable energy, which pursues objectives within the triangle of compatibility with climate and the environment, security of supply and affordability (Bundesregierung 2017). The environmental sector can make significant contributions to the review of set targets by monitoring the environment and provide corresponding information to the public as well as to other stakeholders and decision makers. To this end, observations on the state of the environment must satisfy high objective quality requirements.

A whole chain of processes is involved in bringing observations to the users. It starts with the operation of measuring devices and the recording of measured data and goes through various processing and quality assurance steps up to the presentation of the data. A wide range of requirements and different points of view must be considered, e.g.

- different target groups having special requirements on the data, e.g. easy to understand, completeness, availability in certain technical formats, up-to-date data, access to raw data, free of charge, stability, velocity, data processing, etc.
- operators of measuring networks having an interest in the reduction of expenses, e.g. automated processes, scalable infrastructures, monitoring, etc.
- data providers and users having interests in maintaining quality and provenance of processed data.

The objectives of both, the digitization and the concrete requirements of different parties and stakeholders, must be taken into account. In this article, we would like to focus on technical aspects by sketching an overall architecture, in the awareness that legal and organizational aspects must also be considered and clarified. The presented architecture is intended to cover the entire process from the acquisition of measurement data, through the processing and provision of data up to the visualization. A core requirement for this architecture is that it must be applicable for other application fields, too. All components have to be designed in a generic way.

The load-bearing capacity of the architecture is to be examined in a concrete application, the collection, processing and presentation of air measurement data.

2   Current State and Requirements for a New Solution

The LUBW collects, produces and distributes official environmental information for the federal state of Baden-Württemberg of Germany. In this context, monitoring of air pollution plays an important role and the LUBW is conducting a network of measurement sites tracking meteorology and different air pollutants at least in hourly resolution. The LUBW informs politics and the public in case of exceeded thresholds of pollutants. Accordingly, the data serves as a basis for decision making processes in the Ministry of the Environment and in the domain of public health. In contrast to crowd sourcing data, the LUBW focuses on data quality, consistency and availability.

Atmospheric measurements are presented on the LUBW website using a variety of graphs and tables. The data visualization ranges from ongoing measurements to summarized annual statistics. Unfortunately, the current realization of the website does not benefit from state of the art Web technologies. Graphs and tables are displayed as static images without any interactivity. They are generated via XLST processors and inserted into predefined html pages.

The current workflow relies on an individual, closed data connection, and does not allow establishing Web services which could be used by other platforms. Resulting limitations of the current system are

- data visualization workflow cannot be transferred to other platforms,
- data visualizations are static and not interactive,
- webpages are not optimized for mobile devices,
- maintenance of web pages is hampered by highly individualized data integration processes.

In order to overcome those limitations, a full redesign of the data flow as well as data visualization including the web page generation framework is needed. This framework should include

- a high degree of automation from data ingestion to the visualization processes,
- transferability of services and components to other environmental datasets, e.g. hydrological measurements,
- monitoring and bug tracing functionalities,
- scalability of data services to avoid performance crashes in cases of high user counts,
- increased data availability similar to Open Government Data including REST services and standardized data formats,
- push messages for pollutant warnings on the website and mobile apps.

3 Overall Concept for a Microservice-Oriented Solution for the Handling of Measurement Data

The overall architecture consists of several successive (communicating) phases instead of tiers, a rather horizontal structure with temporal sequences (Fig. 1).

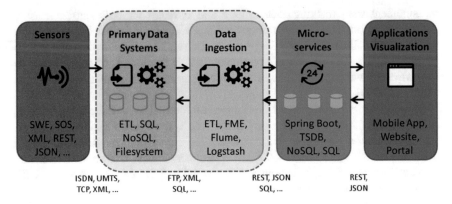

Fig. 1 Overall architecture for the handling of measurement data including core technologies and interfaces

3.1 Sensors

Measuring devices (sensors) on the left generate the original (measured) data. There exists an enormous range of sensors as well as ways for sensor data to enter measuring systems. The reasons for this are different manufacturers, measurement methods, transmission methods, protocols, processes, etc.

For existing measuring systems, interfaces, protocols, data processing and the receiving database systems have been implemented with considerable effort. Changes to them as well as to depending applications are therefore often uneconomical and initially undesirable which has to be considered in the architecture.

More recent sensors do already provide their data via standardized interfaces accessible from the Internet, e.g. via REST or Web services (Sensor Web Enablement,[1] Sensor Observation Service[2]). However, in most cases, direct access to sensors from applications is not possible or useful, e.g. because of limited transmission bandwidths (often using GPRS, UMTS or ISDN), or because of organizational or legal reasons. This means that sensor and measurement data are first transmitted to one or several central data storage systems (primary data systems). Other measurement data are recorded semi-automated or even manually, e.g. for gravimetric measurements of particulate matter (PM10) in laboratories. Such workflows are supported by corresponding specialist applications with human user interfaces.

In the transmission of measurement data from the sensor or measurement system to the primary data systems, a number of possible problems must be solved, e.g. the handling of possible data gaps in the case of disturbed transmission, the conversion of various technical formats, the conversion of units, the mixing of online and

---

[1]http://www.opengeospatial.org/ogc/markets-technologies/swe.

[2]http://www.opengeospatial.org/standards/sos.

offline procedures, etc. Most of them require appropriate logic, e.g. at the interfaces to the primary data systems.

In various ways, these local distributed measurement data are collected and stored on central platforms. We call such existing systems "primary data systems" or "primary databases".

3.2 Primary Data Systems

Primary data systems mainly have the purpose that measured values are being collected and made available in one or more databases, and can be retrieved therefrom and furtherly be processed.[3] They may themselves be complex in structure and include a whole series of processing steps, e.g. conversion of data formats, data (pre-) processing, mechanisms for plausibility checks and quality assurance, aggregation of measured values, etc.

The recording of air measurement data has a long history, many components have grown over time, sometimes very heterogeneously, and the overall system has a high complexity. Measuring systems are built and operated over longer periods of time, mainly because of their considerable costs. ICT systems for recording and providing measurement data have also evolved over time, but compatibility with existing measuring and sensor systems has always been an essential constraint. As a consequence on the one hand measurement systems often do not benefit from latest technologies, but on the other hand existing systems cannot easily be replaced by modern ones.

At LUBW the MEROS system[4] ("Messreihen-Operationssystem Umwelt") is a primary data system of this kind. It bundles time series data from the areas of air quality, nuclear reactor monitoring and floodwater. MEROS is in operation since 1989. Many (specialized) applications rely on the primary data systems and offer possibilities for the use and further processing of data. This is another major constraint that makes architectural changes on existing primary data systems more difficult, since these applications also have to be adapted.

However, since many primary data systems do not meet the requirements of todays distributed, service-oriented systems, additional services are needed, which can serve as a basis for a variety of modern applications. These services are part of the fourth phase "microservices". The third phase ("data ingestion") is responsible for transferring data from the primary data systems to these services.

---

[3]The term "data ingestion" is often used for this process. However, we already use "data ingestion" elsewhere (see below).

[4]http://www.la-na.de/servlet/is/1039/.

3.3   Data Ingestion

Definitions for "data ingestion" vary from "obtaining and importing data for immediate use or storage in a database"[5] to "take data in and put it somewhere it can be accessed".

In our case, we consider data ingestion as the process of collecting and pre-processing data until they are provided for applications using appropriate services. It may well be that these data are already present in further systems (primary data systems) and have been processed and persisted there as well.

In the presented architecture data ingestion is the glue between sensors or primary data systems on the one hand and the microservices on the other, e.g. services for master data, time series, spatial data, metadata, etc. (Braun et al. 2017a) In addition, however, they can perform whole series of other tasks which are still partially implemented in the primary data systems. Therefore, in medium to long term, the two phases of primary data systems and data ingestion will merge.

One of the main objectives of the data ingestion phase is to support data managers in their work and, in particular, to relieve them of standard tasks. Graphically supported methods for the creation and configuration of data flows as well as for the connection of processing steps can be a means for this. This is subject to ongoing studies (Bartosch 2017). In addition, there exist many requirements on tools used in the data ingestion process. They must ensure correct data flow, e.g. aspects like recording and conversion of raw data, data transformation, harmonization of heterogeneous data formats and types, plausibility checks and quality assurance, consistency of data sets, monitoring and error handling, or updating of persisted data have to be tackled. Further requirements and constraints, e.g. performance, scalability, transactions, or exactly-once processing have to be also considered. They are well described in Meehan et al. (2017).

There exists a large selection of powerful tools available on the market which can be used to implement the data ingestion phase on the basis of configurations. Depending on the application, several tools can be used in parallel. In addition to time control, most tools offer an event-based data flow as well as a large selection of ready-to-use conversion, selection and further processing steps.

In the case of the processing of measurement data, the data flow is used unidirectionally, from the sensors towards the consuming applications. However, records may change over time, e.g. if measured values are subsequently corrected by manual interventions within the quality assurance processes, or if data are supplemented with delay, e.g. in case of transmission problems. It must therefore be ensured that the data flow also works for modified or supplemented data sets, and views of the data are consistent at all times and in all phases. The interfaces between adjacent phases have to implement a consistency or coherence protocol, depending on the requirements.

---

[5]http://whatis.techtarget.com/definition/data-ingestion.

In principle, however, also a bidirectional data flow is possible, e.g. when applications produce their own data, e.g. in crowd sourcing applications or edited master data in specialist applications. Depending on the use case, application-generated data is stored in the microservices, respectively their underlying persistence systems, or even transferred back to the primary databases. The assurance of data consistency is currently accomplished by means of time stamps and special flags for marking incomplete, implausible or invalid values. A complete, system-wide framework for ensuring consistency has not yet been implemented.

The data ingestion phase for the processing of air measurement data for the LUBW uses Apache Flume,[6] Elastic Logstash[7] and FME[8] alongside some individually programmed ETL processes. These are operated in the cloud (Google Container Engine).

The primary databases and the data ingestion have central tasks in common. This can cause both phases to grow together in the future, i.e. data ingestion will directly access sensors and write their data directly into microservices, if necessary after having processed them. Technical applications would then no longer operate on the primary data, but directly communicate with the microservices. The possibly different views on the data would have to be adjusted if necessary.

3.4 Generic Microservice Backend (GMB)

In the data ingestion phase, the data is retrieved from the primary data systems and transferred to a service infrastructure. For the provision and dissemination of (measurement) data a set of services is needed and has to be set up. In order to keep efforts on an acceptable level, it is an essential objective to provide the entire information by means of a limited number of generic services (Fig. 2).

For the requirements of a number of environmental portals and mobile applications a total of 8 generic services have been identified: Master Data Service, Schema Service, Time Series Service, (Media and) Digital Asset Service, (Full Text) Search Service, Geo Data Service, Metadata Service and Link Service.

These 8 core services are supplemented by two additional services supporting (and rather belonging to) consuming applications: Application Configuration Service and Data Discovery Service. All services are described in more detail in Braun et al. (2017a) and Schlachter et al. (2017).

All services should preferably be independent from each other. For this requirement corresponds to the component term that is used for microservices (Fowler and Lewis 2016) an implementation using microservices is obvious.

---

[6]https://flume.apache.org.

[7]https://www.elastic.co/de/products/logstash.

[8]https://www.safe.com.

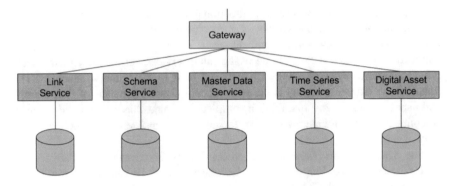

Fig. 2 Most relevant services of the generic microservice backened (GMB)

Packed in runtime containers such as Docker,[9] they can be operated without any additional effort on a variety of possible infrastructures, like dedicated servers, clusters, or in the cloud. Using runtime infrastructures like Kubernetes,[10] operational aspects such as (rolling) updates, monitoring, scalability and load balancing are just a matter of configuration. All services use suitable backend systems, which in particular ensure the persistence of the data. The GMB architecture is abstracted from concrete systems so that these backend systems can be easily exchanged or different backend systems can be used simultaneously. Services provide their functionality through versioned RESTful interfaces via content negotiation. This facilitates the development, maintenance, and replacement of individual services. The architecture provides a messaging infrastructure (channels) for inter-service communication. All services are implemented using the Spring Boot[11] framework.

3.5 Application, Visualization

A whole series of representations exist for measured values and time series. Presentations in the form of diagrams and tables are obvious. However, since measured values always have a geographical reference, representations in the form of maps, e.g. showing measuring stations or as heat maps, can also be applied.

In addition to the selection of a location, the differentiation according to the pollutant or the selection of the temporal reference or aggregation level are standard features. Diagrams must, for example, react to the selection of a substance and display the appropriate time series, the selection of another measuring station must lead to a change of data records, the modification of the time reference to adapt

---

[9]https://www.docker.com.

[10]http://kubernetes.io.

[11]https://projects.spring.io/spring-boot/.

scales and to reload data. Diagrams (and other presentation types) and their mapping to data sets must therefore be parameterizable. The parameterization has to be carried out during the display, e.g. triggered interactively by the user. In addition, there is the requirement to be able to reuse once defined representations, for example, at different locations within a website or a portal.

An answer to this is the Flexvis framework (Braun et al. 2016), which is used to display data, e.g. as diagrams, maps or tables. In detail, it offers the possibility to describe data sources, display types and to define concrete displays as a combination of data source and display type. Both, the definition of data sources and display types can be parameterized, e.g. these parameters can be changed via presets or events. The selection of a specific pollutant by means of pressing a button can thus be carried out e.g. to change the display of a diagram (Fig. 3) or a table. Flexvis offers the integration of almost any display type in the form of web components or JavaScript libraries such as Highcharts or D3.js.

All images are encapsulated as Web components (Braun et al. 2017b). This means a high degree of independence of the display type from the surrounding (Web) application. This makes it easy to use presentations in various CMS and portal systems, but also in hybrid mobile apps. This also means that the display types can adapt to the available size of the display. Responsive layouts and designs are taken into account from the outset even for complex visualization types.

4   First Experiences

The presented architecture contributes to the recording, processing, provision and presentation of air measurement data for the LUBW as defined in the use case.

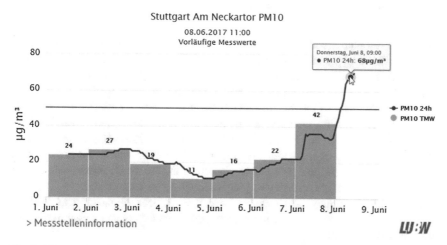

Fig. 3 Diagram for NO$_2$ at the measuring station "Stuttgart Am Neckartor". Details are shown at the mouse pointer position

The MEROS system has not been replaced as a primary data system and continues to serve as data source. So introducing the new architecture for the moment nothing changed in the connection of sensors and measuring systems.

As a consequence, the data ingestion only contains those processes which have not already been implemented in MEROS. Hence, a future merger of the MEROS primary data system and the data ingestion phase will require considerable efforts.

The operation of the data ingestion tools in the Cloud has proven itself for the present use case. Since the target platform GMB is operated in the Google Cloud anyway, a high-performance platform with scalable services and tools is available here. For the massive data provision by the MEROS system, special platforms for data exchange are being used, e.g. a FTP server providing XML files.

Experiences in development and operation of the GMB are very positive, some services are in operation since February 2016. Because of the extensive independence of services a gradual start-up was possible, the actual platform uses microservices providing master data, time series and spatial data. The development of individual services is straightforward and requires relatively short periods of time.

Rising requirements in operation, for example a growing number of accesses, can easily be scaled out on the fly using the horizontal scaling capabilities of the container virtualization infrastructure. The use of containers means extensive independence in the selection of infrastructure operators. The relocation of single or several services is possible.

Within consuming applications the use of a given family of services enables the reuse of generic, highly configurable frontend components including the Flexvis framework and its Highcharts-based diagrams. This additionally simplifies the work of online editors and increases the recognition value for users.

5 Summary and Outlook

With the presented architecture, a complete processing chain from the sensor to the visualization in applications could be successfully established using the use case "air measurements in Baden-Württemberg". A whole range of tools and technologies were used, some out of the box, others newly developed.

Thus, a key point of the federal government's digitization initiative was taken up and implemented. With the provision of measurement data, e.g. PM10 and NO_2, it is possible to check the impact and effectiveness of measures in the field of renewable energies or transport. An important pillar for this is the provision of measurement data and understandable visualizations for measurement data for the public. However, a whole series of points remain to be worked on:

The existing primary data systems have not yet been replaced. This leads in many cases to redundant data maintenance, additional efforts and sometimes issues with data consistency.

Many applications continue to access the primary data systems directly. These would have to be converted to the Generic Microservice Backend.

The consistency or coherence of data has to be ensured over all phases, e.g. using a coherence protocol.

The transferability of the architecture to other application fields must be checked, in particular for deviating boundary conditions such as tougher real-time requirements, for smaller time scales, e.g. in the field of energy systems.

All in all, the new architecture provides a useful toolbox for the processing and visualization of measurement data. It is to be applied shortly to flood data and other data in the area of running waters.

References

Bartosch B (2017) A new generic approach to data ingestion and quality management in microservice based data management environments. Diploma thesis, work in progress, KIT, to be published in summer 2017

Braun E et al (2016) Generic web framework for environmental data visualization. In: Advances and new trends in environmental informatics EnviroInfo 2016, Berlin, Germany, 14–16 September 2016, pp 289–299. doi:10.1007/978-3-319-44711-7_23

Braun E et al (2017a) A generic microservice architecture for environmental data management. In: International symposium on environmental software systems (ISESS), Zadar, Croatia, 10–12 May 2017

Braun E et al (2017b) A lightweight web components framework for accessing generic data services in environmental information systems, submitted to EnviroInfo 2017, Luxembourg

Bundesregierung (2017) Energiewende—Schutz für Umwelt und Klima. https://www.bundesregierung.de/Webs/Breg/DE/Themen/Energiewende/Energiesparen/stromnetze/_node.html. Accessed 25 April 2017

Fowler M, Lewis J (2016) Microservices—definition of this new architectural term. http://martinfowler.com/articles/microservices.html. Accessed 31 Aug 2016

Meehan J, Aslantas C, Zdonik S, Tatbul N, Du J (2017) Data ingestion for the connected world. In: CIDR 2017 8th biennial conference on innovative data systems research; chaminade, CA, USA, 8–11 January 2017. http://cidrdb.org/cidr2017/papers/p124-meehan-cidr17.pdf

Schlachter T et al (2017) A generic web cache infrastructure for the provision of multifarious environmental data. In: International symposium on environmental software systems (ISESS), Zadar, Croatia, 10–12 May 2017

A Data Context and Architecture for Automotive Recycling

Clayton Burger and Alexandra Pehlken

Abstract Automotive environmental and financial perspectives form important part pillars of the sustainability of the automotive sector. Coupled with increased pressure on automotive recycling quotas in various countries and constrained limits on various critical metals due to availability, sourcing and recyclability, a clear need for data-driven decision support arises. This paper investigates such a data-driven architecture in the context of European automotive recycling. The presented architecture is grounded in the context of reporting requirements, stakeholder interests and the legal landscape in the European Union. Various key data aspects are presented in terms of resource, market, environmental and assembly data sources. The availability of such publicly available data is a pervasive problem in this research domain with strategies for sourcing of such data throughout the production, use and recycling lifespan of the automobile presented. Key partners, such as recyclers and dismantlers, are included in the development of the architecture to ensure appropriate data sources are included and mappings between different data granularities are suitably accounted for in a data schema. Implementation and adoption strategies are briefly outlined together with recommendations for governmental support to encourage OEM participation and responsibility.

Keywords Cascading · Automotive recycling · Material efficiency · Automotive assembly

C. Burger (✉) · A. Pehlken
Cascade Use, Carl von Ossietzky University of Oldenburg, Oldenburg, Germany
e-mail: clayton.burger@uol.de

A. Pehlken
e-mail: alexandra.pehlken@uol.de

© Springer International Publishing AG 2018
B. Otjacques et al. (eds.), *From Science to Society*, Progress in IS,
DOI 10.1007/978-3-319-65687-8_19

1   Introduction

Automotive recycling is a domain that has garnered interest in recent years due to environmental pressures of car production (Elwert et al. 2015; Pehlken and Kalverkamp 2015; Andersson et al. 2017). With the proportional increase in vehicle production and use, the environmental impact has grown substantially larger, specifically with emissions during production and usage, energy demands and material requirements (Pehlken et al. 2014). Focusing on material efficiency, the EU has placed various rising legislative quotas and requirements on countries to increase recycling efficiency and efforts (first Directive 2000/53/EC, now in its 6th revision 2016/774/EU). 95% of materials in end-of-life vehicles must be recovered, while the remainder is considered waste, including a large amount of critical metals, many of which are not considered recoverable after alloying with other metals in the shredding and recycling phases (Elwert et al. 2015). This loss of materials is notable as these metals are scarce in supply and are steadily increasing in demand due to their applications in electronics, such as neodymium in permanent magnets in storage media in car control systems and cobalt in many batteries.

The environmental pressure to improve car recycling is a global problem with a large set of challenges (Chen and Wang 2014; Venkatesan and Annamalai 2017). One of the foremost challenges in recycling are the consumption patterns of consumers as they form the demand for vehicles. The waste hierarchy outlines different actions with increasing environmental impact, with reuse requiring the minimum in energy and material inputs after prevention considerations. Recycling forms the next step, followed by energetic recovery and disposal. Encouraging consumers to reuse, when possible, and recycling, where necessary, forms the foundation of a behaviour where materials can be conserved to minimise environmental impact. This paper thus investigates how information systems can be applied to this problem through the development of an architecture to engage various stakeholders with decision support to encourage reuse and recycling by displaying the composition of a car in terms of materials, components and the associated reuse and recycling value potentials.

This paper is structured in terms of the research objectives and methodology employed to develop the architecture proposed Sect. 2, followed by a short discussion of stakeholder requirements and data provisions Sect. 3. The architecture is proposed Sect. 4 with implementation and adoption considerations, followed by a discussion of observations and future research avenues.

2   Research Objectives and Methodology

Automotive recycling is a topic that has garnered interest worldwide due to environmental pressure, but research into data generation and management in the domain has been limited (Andersson et al. 2017). The main research objective of this paper is therefore: *"To propose and contextualise an architecture for automotive recycling data support for various stakeholders that supports effective environmental decision support."* In order to generate and verify such an architecture, three secondary objectives must be addressed, namely:

- Identify the decision support data requirements of recycler, dismantler, governmental and public stakeholders;
- Propose a candidate architecture that incorporates feasible data requirements of stakeholders as component data streams; and
- Propose an implementation strategy for realising the theoretical architecture as a software framework.

In order to sufficiently address the posed research objectives, and finally the main research objective, with developed artefacts (identified requirements, potential data sources and the assembled architecture), this study employs a mixed method approach with deductive assertions of requirements with inductive generation of the architecture. This mixed method approach is facilitated by using the Design Science Research (DSR) methodology due to its applicability in information systems studies that consolidate public user requirements with appropriately generated artefacts (Hevner et al. 2004; Hevner 2007). The prescribed process requires a careful and rigorous study of extant systems and solutions, relevant literature and profiling of stakeholders to develop a comprehensive overview of the domain under investigation. By examining the context and state-of-the-art in automotive recycling, this process is fulfilled while concurrently employing the cyclic and iterative nature of the methodology to ensure rigour and relevance, as prescribed by Hevner (2007) to design and generate the identified artefacts. A set of theoretical guidelines for implementation are concurrently generated through these cycles.

3 Context of Automotive Recycling

Considering cars as the focus of automotive recycling, four major stakeholder groups are identified. These stakeholders are discussed followed by a short discussion of the current data landscape in car recycling.

3.1 Stakeholder Interests

Four key groups of stakeholders were identified as drivers for consumer behaviour and impacting on material conservation. These stakeholders are: public consumers, governmental stakeholders, recycling stakeholders and dismantling stakeholders. Subsets of **public consumers** have shown to have an increasing interest in being environmentally responsible (Venkatesan and Annamalai 2017), with financial interests in maximising savings. **Governmental stakeholders** are concerned with monitoring material inputs and outputs of countries, especially when critical metal prices are unstable and availability is an increasing concern (Elwert et al. 2015). The remaining two groups, **recyclers and dismantlers**, have direct interests in the value and recoverability of materials and components respectively. Both of these stakeholders benefit from information about the composition of cars and components, but this information is typically not publically available and impacts on their business models. All four of these stakeholder groups pose various queries about car and component compositions, environmental impact, financial worth and reusability.

3.2 Data Availability

Data sources for recycling are a very prevalent issue in decision making. The data landscape is extremely limited due to OEMs not providing information about material inputs and component/car compositions due to company secrets.

A number of questions arise when investigating the available data sources, namely with regards to resource information, market information, environmental information and disassembly information. These issues, together with identified candidate public data sources, are illustrated in Fig. 1 for brevity. A data-driven architecture requires relevant and feasible data inputs, which are very limited in supply. For the architecture posed in this study, various data sources are identified that require automatic parsing or web service wrapping, which increases development and maintenance complexity. Reusability of these services through modular development thus forms one of the design considerations of the architecture. By considering the major categories of data sources, the schema developed in an implementation of the decision support architecture can thus be interchanged with country-specific or additional data inputs.

4 Development of Software Architecture for Automotive Recycling

Developing a suitable architecture for automotive recycling decision support requires consideration of various concepts in order to consolidate the stakeholders, reporting requirements and data sources available. The process of inductively developing the architecture is outlined, followed by an analysis of the assembly process of generating usable data streams and then connecting them to fulfil the stakeholder goals of the architecture, resulting in the proposed architecture as an artefact. The application of information systems best practices is also outlined with motivations for the architectural patterns included in the artefact.

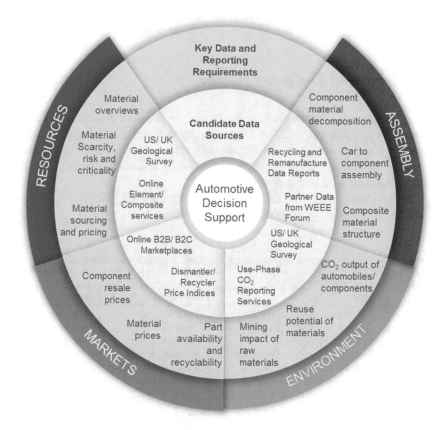

Fig. 1 Overview of high-level data requirements and candidate data sources, categorised by reporting perspective

4.1 Architecture Development Process

The development of the recycling data management architecture requires foresight of the implementation potentials and is thus aligned with IT infrastructure in mind, based on the guidelines posed by Hohpe and Bobby (2004). Planning the data inputs and pipeline processes with reporting goals in mind assist in ensuring that redundant data is not added and that relevant data is not excluded before the architecture is implemented as a framework. The process employed to develop an architecture typically follows the standard information systems development process of iteratively identifying stakeholders, requirements, goals, data consolidations and reporting results, which is aligned with the DSR process used in this study. The main pillars of the design are outlined by the user requirements (Sect. 3.1) and data availability (Sect. 3.2). Assembling the so-called pillars into data flows requires data mappings, especially in the domain of automotive recycling due to the extremely heterogeneous nature of the available data inputs and the transformations required to generate usable data streams from these sources.

4.2   Granularity and Query Support

The architecture must support various queries to facilitate reasonable decision support through data provisioning. The data requirements Fig. 1 form a baseline mapping of the queries provided by the architecture. Three levels of granularity emerge from the stakeholder requirements and data availability, namely at a *car level*, *component level* and *material level*.

The relevant queries and their stakeholder interests are outlined in Table 1, which forms a key perspective of the architecture.

Table 1 Mappings of focussed stakeholder requirements to architecture supported queries

Stakeholder	Granularity		
	Car	Component	Material
Public consumers	- Environmental impact in CO_2 - Composition of car components for substitution -Composition of car in respect to material criticality, risk and overall value -Composition of car in respect to overall component value	- Material composition and value of a component - Public market value of component	- Material criticality and risk assessment
Government	- Environmental impact in CO_2 -Composition of car in respect to material criticality, risk and overall value	- Material composition and value of a component	- Public material price - Material criticality and risk assessment - Safety of material extraction
Recyclers	-Composition of car in respect to material criticality, risk and overall value -Composition of car in respect to overall component value	- Material composition and value of a component - Public market value of component	- Public material price - Material criticality and risk assessment - Safety of material extraction
Dismantlers	- Composition of car components for substitution -Composition of car in respect to material criticality, risk and overall value -Composition of car in respect to overall component value	- Material composition and value of a component - Public market value of component	- Public material price - Safety of material extraction

4.3 Heterogeneous Data Inputs and Mappings

A candidate architecture prototype was developed from the stakeholder requirements as an integration platform for the data services required. The architecture Fig. 2 is composed of three layers with clear separation of concern.

The *presentation layer* is responsible for gathering user inputs in a structured fashion for the other layers through service packaging, while also generating reports from queries generated using a querying protocol to support the queries outlined in Table 1.

The *decision support* layer bridges the interface with data services by structuring queries in respect to the environmental/financial decision features to show which is more favourable with respect to the waste hierarchy and the query considered. Analytics algorithms can be added to further increase the query power of the architecture with meta-data and meta-analysis.

The *data access layer* connects the other layers to a relational database which uses traditional data transformation, extraction and loading to populate and maintain a local repository of the heterogeneous data services with components that support these operations. The populator component coordinates three interfaces with the query broker and mapping engine at the three granularities.

The material data interface Fig. 3 connects online element information, parsed geological survey data and parsed web price information of various raw materials and identified composites. The geological survey and market information is directly parsed and frequently updated in the RDMS.

The component data interface Fig. 4 connects data from third party partners that have reported on laboratory analyses of components in terms of their material inputs and recycling output fractions. A candidate data stream for this service comes from European Union funded projects and published reports which should be parsed into the framework implementation. Similarly, market price values of components are to be parsed in real-time from market communities, such as EBay.

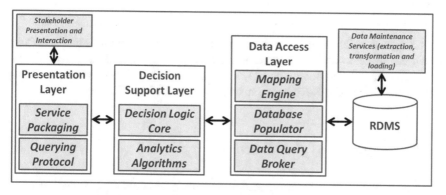

Fig. 2 Candidate architecture with modular units and black-box relational database

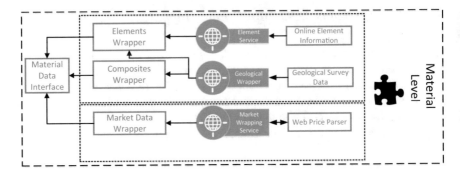

Fig. 3 Material data interface

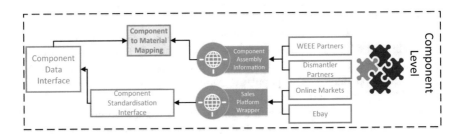

Fig. 4 Component data interface

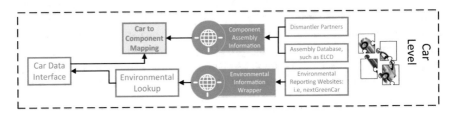

Fig. 5 Car data interface

The car data interface Fig. 5 finally connects data of the bill-of-components for each car made available by dismantling third parties, as well as provides environmental lookups to determine CO_2 impacts of cars from governmental reporting websites, such as the NextGreenCar service in the United Kingdom.

4.4   Integration of Best Practices

Use of well-defined role encapsulation and minimal coupling ensures interoperability of key components of the architecture, both at a tier level and at a realised class level (Lytra et al. 2013). Encapsulation is encouraged in the posed architecture through clear responsibilities for each component, such as the database populator which is composed of various sub-services for each data granularity.

Adoption of successfully demonstrated architectural design patterns, such as the model-view-controller (MVC), multitier design (N-Tier), layer-based design and implicit invocation patterns have been demonstrated in information system development to increase framework adoption and pervasiveness though strong design using best practices (Oduor et al. 2014). A clearly defined layer architecture that employs MVC design practices is thus ideal for transformation of the architecture into an implemented framework. The layer design is inherently included due to the separation of concern employed and furthermore through the granularities identified. By realising the architecture as it has been posed, a distributed or cloud-based implementation is possible, which further increases the reusability of the architecture.

5 Implementation Considerations and Adoption Strategies

The realisation of the identified architecture as a software framework inherently has various key considerations. The target platform is a web-based system that offers mobile support, either directly as a native application or through responsive design as a web-based application. Due to the current development landscape, the ideal implementation platform would natively support MVC development, such as ASP. net 4.0, Java MVC or Ruby-on-rails with PHP, amongst other current offerings. Another key input is expected product longevity, suggesting an open-source platform with public source, as the data contains no confidential or private information.

Adoption strategies to consider should include the maintenance of the software framework, the iterative development of the framework and legal adoption. The market presence of this architecture realised as a software tool would be limited due to the fact that no private or OEM information is included, thus an open-source platform where public stakeholders can contribute and maintain the system would be ideal. An alternative would be to incorporate marketing sales or to offer the service as part of an automotive environmental consultancy. These factors affect the iterative development of the architecture and have similar arguments. Finally, the legal adoption of the architecture at an EU or national level would encourage maintenance of data sources, potential addition of new sources and possibly even direct OEM support. The legislative potential falls outside the scope of this study.

6 Conclusions and Recommendations

This paper proposed a high-level architecture for automotive environmental and financial decision support as part of the goal of increasing material efficiency and reducing the environmental burden of a car's life cycle. Various limits frame this study, such as the decision to focus only on whole cars rather than including the car parts and the decision to maintain full transparency by not including proprietary information from OEMs. This contribution thus forms part of a inter-disciplinary project focusing on these topics from various perspectives, with this paper showing the information systems perspective. The development of this architecture highlights various key findings, namely the stakeholder identification, data sources available and finally the potential implementation of the architecture itself. Further user acceptance and impact research is required to fully investigate the potential of the architecture when realised as a software framework, which forms the future research goal of this study.

Acknowledgements The research for this paper was financially supported by the German Federal Ministry of Education and Research (Grant no.: 01LN1310A). The authors gratefully acknowledge their support. The content is solely the responsibility of the authors and does not necessarily represent the official views of the ministry.

References

Andersson M, Söderman ML, Sandén BA (2017) Lessons from a century of innovating car recycling value chains

Chen M, Wang J (2014) Recycling of End-of-Life vehicle: regulation, management and prospects. J Shanghai Jiaotong Univ 1:22

Elwert T, Goldmann D, Römer F et al (2015) Current developments and challenges in the recycling of key components of (hybrid) electric vehicles. Recycling 1:25–60

Hevner AR (2007) A three cycle view of design science research. Scand J Inf Syst 19:4

Hevner AR, March ST, Park J, Ram S (2004) Design science in information systems research. MIS Q 28:75–105

Hohpe G, Woolf B (2004) Enterprise integration patterns: Designing, building, and deploying messaging solutions. Addison-Wesley Professional

Lytra I, Tran H, Zdun U (2013) Supporting consistency between architectural design decisions and component models through reusable architectural knowledge transformations. In: European Conference on Software Architecture. pp 224–239

Oduor M, Alahäivälä T, Oinas-Kukkonen H (2014) Persuasive software design patterns for social influence. Pers Ubiquitous Comput 18:1689–1704

Pehlken A, Kaerger W, Chen M, Mueller D (2014) Environmental issues in automotive industry. Springer, Berlin, Heidelberg

Pehlken A, Kalverkamp M (2015) Kaskadennutzung im Automobil: Realität oder Zukunftsmusik. Recycl und Rohstoffe 174–181

Venkatesan M, Annamalai VE (2017) An Institutional Framework to Address End-of-Life Vehicle Recycling Problem in India. In: SAE Technical Paper. SAE International

Application of Methods of Artificial Intelligence for Sustainable Production of Manufacturing Companies

Martina Willenbacher, Christian Kunisch and Volker Wohlgemuth

Abstract An energy- and resource-friendly production is an important key performance indicator for industrial companies to work economically and thus remain competitive. For this software systems are necessary for analysis, evaluation, diagnosis and planning. Thanks to intensive research efforts in the field of artificial intelligence (AI) a number of AI based techniques such as machine learning, deep learning and artificial neural networks (ANN) have already been established in industry, business and society. In this paper, we address the problem of energy- and resource efficiency in production processes of manufacturing companies. We present an approach to improve energy- and resource efficiency by methods of AI. We propose an in-progress idea to extend the possibilities of using methods of AI for optimizing material and energy flows. In addition to processing the process data, an integrated database of measures is designed to support sustainable production. The investigations are carried out prototypically using an ANN in combination with fuzzy logic and evolutionary algorithms (EA).

Keywords Artificial intelligence · Neural network · Machine learning · Energy efficiency · Material- and energy flows

M. Willenbacher · C. Kunisch · V. Wohlgemuth (✉)
HTW Berlin, Department Engineering-Technology and Life,
Environmental Informatics, Wilhelminenhofstraße 75A, 12459 Berlin, Germany
e-mail: volker.wohlgemuth@htw-berlin.de

M. Willenbacher
e-mail: martina.willenbacher@htw-berlin.de

C. Kunisch
e-mail: christian.kunisch@student.htw-berlin.de

© Springer International Publishing AG 2018
B. Otjacques et al. (eds.), *From Science to Society*, Progress in IS,
DOI 10.1007/978-3-319-65687-8_20

1 Motivation and Goal

Due to the constantly increasing quantity and complexity of the data to be processed, the predominantly used conventional mathematical methods are only partially suitable for mapping and computation of holistic and interdisciplinary processes and making decisions and predictions on this basis. Also dealing with the presence of sufficient problem knowledge as well as the processing of blurred knowledge and erroneous data cannot be solved by conventional methods (Page et al. 1995). One way to circumvent these problems is, for example, the use of AI methods. The development leaps to research in the field of AI are becoming larger and follow at shorter intervals. More and more companies are investing more and more sums in the development of their own AI applications or in takeovers from companies that are involved in the development of AI applications.

The example DeepCoder shows how important the current advances in research and science are about AI. DeepCoder is an AI developed by Microsoft Research and the University of Cambridge, which is able to solve programming problems by reusing code lines from other programs. The research presented at the International Conference on Learning Representations 2017 describes how an ANN is trained by means of deep learning so that it is able to observe input and expected program output as well as predefined code in the course of the automatic program synthesis (Balog et al. 2017). The AI is therefore in the position of what many scientists have dreamed so far: programs develop programs. AI is also referred to as Computational Intelligence. The purpose of this research is to imitate or simulate behaviors of human experts and optimizing processes from nature on a computer. The goal is to simulate associative and creative thinking and to verify with the computer (Grauel 2008). The aim of this paper is to demonstrate the possibilities of using AI methods to analyze and optimize processes. Furthermore, it is to be investigated whether a surplus value for the use of AI exists against conventional methods for the analysis, calculation and evaluation in environmental management systems and how this results. In addition to the development effort and the performance of the ANN, the analysis of the impact should also take into account aspects such as user friendliness and the amortization of the total expenses as a function of long-term use. In addition, the ANN, which is to be developed, should also consider and investigate factors that are difficult to reproduce, such as the aging and utilization behavior of machines and systems, as well as the associated change in energy consumption. This also includes the conceptual design of an intelligent automation of the creation of the LCA, through which the relevant company data are provided. At the same time the measured and calculated data of the material balance sheet should be recorded during the production process and estimates for the holistic depiction of the production process should be made using methods of AI. In addition to energy, raw materials, operational materials and physical quantities (e.g. emissions, temperature, pressure, phon/sone) products, coproducts, waste and other environmental aspects are also considered. In addition, measures for the optimization of energy

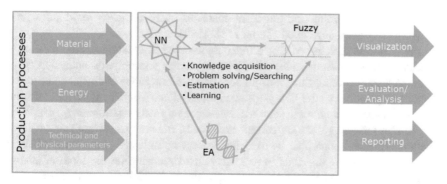

Fig. 1 Applied methods of AI (Willenbacher 2017)

and material flows as well as risks within production are to be identified and used for the impact assessment. The key research questions addressed in this paper are:

- Which methods of AI can support existing environmental management systems in their calculations, evaluations and knowledge bases?
- Is there an advantage over conventional methods in the use of AI methods for optimizing processes with respect to environmental sustainability, and how is this also designed with regard to the adaptability of existing systems and long-term use?
- Can intelligent methods be derived by means of AI methods and these can be improved by self-learning methods and adapted to new and changing processes? (Fig. 1)

2 State of Science and Technology

As a result of the highly dynamic development processes (economic, demographic, sociological, ecological and biological processes), high demands are placed on the finding of scientifically founded answers to environment-specific questions and the formulation of models. Decision-making is made difficult on the one hand by the shortness of the time available for the causation of solutions and by the complexity of the systems and processes of the environment.

In studies of environmental systems, the large amount of data is a further problem. They are usually dependent on space and time and are collected and stored in geographic information systems, environmental information systems or environmental databases. A knowledge-based system was linked to CONKAT (CONnectionistic Knowledge Acquisition Tool) with a connectivist model. In this case, data set structures are recorded by a neural network and provided as symbolic knowledge (Ultsch and Halmans 1991). In the area of emission reduction, a methodology for production tuning mechanisms based on fuzzy logic and neural networks has been developed (Tuma et al. 1993).

Furthermore, there is work on a fuzzy logic-based forecasting system for energy consumption (Lau et al. 2008) as well as an energy consumption analysis system with evolutionary algorithms. For process optimization in the chemical and process engineering industry, the company atlan-tec Systems GmbH offers the software NeuroModel® (http://www.atlan-tec.de/loesungen/prozessoptimierer/), which is aimed at minimizing costs. A current example of the company Google shows that methods of the AI can also be used for process optimization for a more efficient energy consumption and thus have a relevant relationship on the topic of energy and material flow management. An energy and material flow management is the systematic approach to analyze and optimize energy and material flows in companies. The main goal is to reduce the energy used in a company as far as possible (Junker et al. 2015). At the Start Up Energy Transition Tech Festival (TechFestival 2017), which was held in Berlin in March 2017, the focus was on solutions that can be used to achieve more energy efficiency through technology (DeepMind 2017). It shows how the cost of cooling in Google's data centers has been reduced by 40% by AI DeepMind. Google has entrusted the AI DeepMind with the control of cooling in its own data centers. The goal was to train the system further and possibly find improvement potential. The AI was left control of the cooling systems, the fans as well as the windows and almost 120 other things (DeepMind2 2017).

It analyzed historical and current data from 1,200 different datapoints of sensors within the server racks and reached a savings of several hundred million dollars per year by automatically adjusting the cooling schedules and increasing efficiency. The following graph shows how machine learning has reduced energy consumption in Google's data centers. At the point "ML Control On" the analysis and the optimization of the AI started and at the point "ML Control Off" it stopped. The graph falls to a significant lower Power Usage Effectiveness (PUE) value and rises again to a higher PUE value. The PUE value indicates how efficiently a data center uses energy. It is the ratio of total amount of energy used by a data center facility to the energy delivered to computing equipment (Fig. 2).

This optimization reduced the total energy consumption of Google's data centers by 15% (DeepMind3 2017). The purchase of DeepMind by Google in 2014 for around 600 million US dollars will have paid for itself in the shortest time (DeepMind4 2017).

Another example that shows the cost-saving potential by the use of AI systems is the Japanese insurance company Fukoku Mutual Life Insurance. The company introduces an AI based on IBM's Watson system that evaluates medical and other documents to calculate payments due. The AI system is to replace 34 of the 131 employees of the claims department. The cost of setting up the AI system amounts to around 1.6 million Euros and for the operation to approximately 120,000 Euros annually. The annual personnel costs are around 1.1 million Euros (Watson 2017). Accordingly, the costs have already paid for itself two years after the commissioning of the AI system.

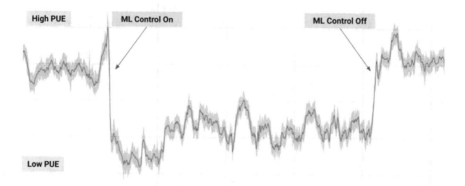

Fig. 2 AI DeepMind reduces energy costs in data centers (https://deepmind.com/blog/deepmind-ai-reduces-google-data-centre-cooling-bill-40/)

Furthermore, ANNs are applied in the modeling of ecological data and are used for the evaluation of water quality measurements in streams (Werner 1996). ANNs are also used to optimize industrial processes with the aim of environmental pollution relief (Tuma and Haasis 1994) and to smog prediction (Ultsch and Beiersdorf 1994). An AI solution in the area of process optimization on the basis of the data of substance and energy streams with consideration of the difficult to represent factors such as the aging and utilization behavior of machines and plants as well as the associated change of the energy consumption is hitherto not known. In the following the methods of AI to be used are explained briefly before the technical implementation is considered.

2.1 Artificial Neural Networks

The process optimization for an efficient material and energy consumption in the production is an essential component in the field of operational environmental informatics. This very complex area is particularly characterized by a very high number of parameters, which are often not directly related. In the modeling and simulation of energy and substance flows, connections between the input and target parameters, which are nonlinear, can occur. In order to provide a meaningful analysis or prognosis, the conventional mathematical-statistical modeling methods are not sufficient. ANNs are also able to detect non-linear relationships between the investigated parameters. For example, the energy consumption of a machine in production can increase exponentially as the quantity produced is increased. The amount of waste in comparison to the production amount can also be in a non-linear relationship. In order to uncover optimization potentials, ANNs are thus offered which more compactly represent multi-dimensional, non-linear connections and simplify their evaluation (Willenbacher 2011). For the ANN to be developed described in this article, a non-linear activation function of the output layer is therefore selected. As a learning method, the supervised learning is used to adapt the network configuration on the basis of a target-actual comparison, as well as the reinforcing learning, which gives the system the option of decision-making based on previous training data. The challenge lies in the combination of different AI methods with the ANN in order to achieve optimal solution results in different problem situations.

2.2 Fuzzy Logic

Fuzzy systems are an extension of classical logic in terms of unspecified quantity affiliation and rules for which there are no exact regulations. They not only process values such as 1 and 0, but also intermediate values (truth values), e.g. 0.5, so that fuzzy data like A BIT, SOMETHING, GOOD or STRONG can be treated mathematically. It is crucial that no simple mathematical (linear) function can be found behind this logic. Rather, the decisive values must be gained from experience, observations and empirical investigations.

In order to eliminate the disadvantage of the black-box view of the ANN and to enable knowledge integration into the network, the ANN to be developed is combined with fuzzy sets. This serves as a preselection of decision making for the reduction of complexity and computing power. The weighting of the ANN takes place via fuzzy sets. Thus the so-called degrees of truth can be assigned to the rules defined for the ANN. This means that the same state triggers different actions. An example of this method is illustrated in the following sketch model.

As an example, the intervals A = [10, 50] generated waste in tons and P = [100, 1000] produced quantity in pieces are used. On the basis of the generated rules, the degree of fulfillment is subsequently determined via the membership function in order to fuzzify the system (Table 1).

Table 1 Example for a fuzzy set

Rule	State 1 "Waste generated"	State 2 "Produced amount"	Action
1	Little waste generation	Little production	Action 2
2	Medium waste generation	Little production	Action 3
3	Much waste generation	Little production	Action 3
4	Little waste generation	Medium production	Action 1
5	Medium waste generation	Medium production	Action 2
6	Much waste generation	Medium production	Action 3
7	Little waste generation	Much production	Action 1
8	Medium waste generation	Much production	Action 2
9	Much waste generation	Much production	Action 3

2.3 Evolutionary Algorithms (EA) for the Optimization of Production Planning and Process Control

Evolutionary algorithms are a class of methods for finding an approximation solution for an existing optimization problem. These methods belonging to the class of metaheuristics are based on the biological principles of mutation, reproduction and selection. The basis for this is the propagation or evolutionary process. Through mutation, the genetic heirs are able to adapt to the changing environmental and living processes. The best individuals adapted to the environment replace the weaker and pass the optimized genetic information to the next generation (Lippe 2006). In the 1960s and 1970s, the first areas of use of EA were founded in the areas of thermodynamics and structural engineering.

The aim of this research is to examine the extent to which supply chain management, i.e. the processes in a company along the supply chain/delivery network, can be optimized with EA with regard to the material and energy flows, taking into account existing information flows. For example, the order of the produced goods can be considered as well as the logistical supply of components. The challenge lies in the consideration of conflicting criteria. The objective is to find the element of the search space, which provides the best value for all target functions. The following points are to be considered:

- The production of several products of a plant with the aim of determining the order of the best possible production quantity under optimum utilization of the plant;
- The assessment of production facilities against the background of resource efficiency and energy consumption, taking into account various factors (cost, production increase, maintenance, repair, useful life, throughput);
- The assessment of the production of secondary products against the background of resource efficiency and energy consumption taking into account various factors (profit increase, plant utilization, personnel expenses).

A solution approach is to get as many pareto-optimal solutions as possible for multiobjective problems. For the calculation the algorithms Strength Pareto Evolutionary Algorithm (SPEA2) (Zitzler et al. 2001) and Pareto Archived Evolutionary Strategy (PAES) (Knowles et al. 1999) were used. With SPEA2 (Kruse et al. 2015) a finer fitness allocation is considered, which takes the number of the dominant solutions of a problem solution into consideration as well as an improvement of the efficiency of the diversity calculation by the determination of new neighbors (density measure). Non-dominant individuals (Pareto limits) can be stored in an archive, which is characterized by an archive size limitation. This can only add new ones by removing older individuals (Mehnen 2004). In the PEAS, the archive is presented as a multi-dimensional hash table in which the hash values define the niches (Kruse et al. 2015).

3 Implementation

An ANN is to be designed and implemented in combination with methods of the AI, which, on the basis of data of modeled substance and energy streams as well as other data of the company (measures catalogs, controlling data etc.), identifies the different parameters of the company's environmental management and determines optimization potentials within the running processes. These optimizations provide the causes and effects of varying process parameters, which should provide action derivations. The resultant database of measures thus becomes an adaptive decision support (Fig. 3).

For this, in a first step possible fields of application in the manufacturing industry are identified and the technical requirements are analyzed in order to develop various practical applications and meaningful extension conditions for the integration of further indirect influencing factors. The advantage of the use of ANN in the optimization is that no explicit knowledge about the actual solution is necessary as the network "learns" the solution process by means of training data. An attempt should be made to calculate nonlinear functions and dynamic systems by means of methods of the AI and thus to take into account difficult factors such as aging, usage behavior and the associated change in resource dependencies. Therefore, the ANN is not viewed separately, but is combined with methods of computerized thought processes, logical closing and intelligent interactions of artificial machines.

In the second step, the identified processes with their complex problems are modeled by natural-analogous procedures. For this purpose, data sources have to be defined which, in terms of their quality and quantity, enable a meaningful training of the network. After the selection of the combination of the network architectures as well as the determination of the influence variables (input/output) and learning rules, the modeling and programming of the network is done. The results obtained during the training phase for the classification of data from various sources, which are partly not directly related, are intended to represent the relationship between the impact and the measure. By self-learning methods of the applied AI methods, these measures are continually adapted and extended with regard to the sustainability of

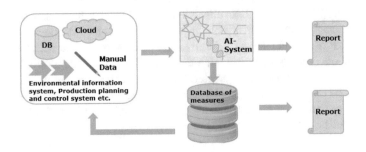

Fig. 3 Conception (Willenbacher 2017)

Table 2 Exemplary representation of the data entering the modeling

	Data
Processes	Manufacturing, transport, storage
Input	Energy, water, raw materials, auxiliary materials, supplies
Output	Waste materials, residues, hazardous materials, energy, waste water, emissions, final products, co-products, residual materials, energy, emissions
Connected environmental aspects	Emission, noise, storage, machine maintenance, resources for production facilities, packaging
Weak aspects	Usage behavior of the machines (downtimes, workload, maintenance intervals), performance characteristics of the machines, aging, ambient temperature, quality assurance, personnel

production processes, so that a practice-oriented database of measures is available. The following table shows the data that is used in the modeling (Table 2).

The development of the ANN with fuzzy logic and evolutionary algorithms is carried out in the programming language C/C++ with Microsoft Visual Studio. The software MemBrain is used for the simulation of the network. MemBrain provides a DLL for linking to Visual Studio, which can be used with the languages C, C++ and C#. The data evaluation is carried out with the software R and the associated package nnet. The database of measures will be implemented as web application, developed with MS SQL in Visual Studio. By implementation of a XML interface the possibility of transfer of external data from existing databases of measures should be created. Bayesian networks are to be used to derive the appropriate measure. Bayesian networks are directed graphs, which allow conclusions to be drawn on the effect of certain causes. Thus, possible consequences can be recognized even before the occurrence of the undesired event (e.g. machine failure) and if necessary measures can be taken. The graphic below shows as an example a BN for native fish in the Goulburn Catchment, Victoria (Pollino et al. 2007b) (Fig. 4).

4 Outlooks

With ANNs, it is also possible to make statements on the behavior of a system if the data basis is incomplete. Likewise, they can achieve a good meaning with slightly falsified inputs. Especially in the field of interdisciplinary data sets and dynamic models, as is often the case in operational environmental informatics, the use of ANN in combination with fuzzy logic and EA offers significant advantages:

- Learning ability of the ANN from training data as well as associative ability, i.e. Solve similar problems after training
- High fault tolerance of the ANN against noisy data
- Illustration of human inference by fuzzy systems

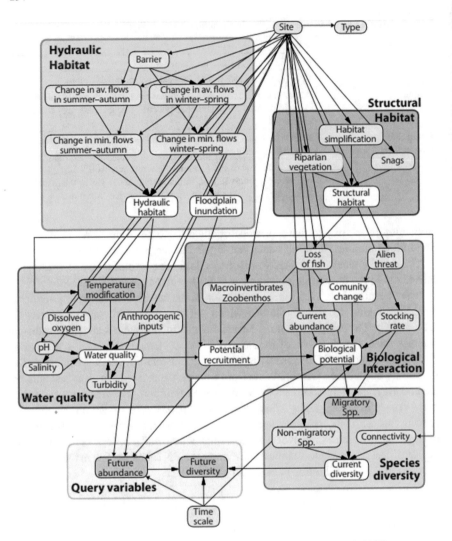

Fig. 4 BN for native fish in the Goulburn Catchment, Victoria (Pollino et al. 2007)

- Decision support for first-time tasks through the operationalization of linguistic, subjective values by fuzzy systems
- Decision support for problems with many solutions or problems without enough structural knowledge by EA

The disadvantage, however, is that knowledge within ANN is difficult to interpret. This is particularly the case when analyzing the system behavior with incomplete or partly erroneous data, since these results are often difficult to comprehend. The flexibility in the design of the ANN also has disadvantages. There is no general method for the design of ANN. Therefore the development is mostly experimental and very complex, combined with high development costs. However, it must be noted that when developed, ANNs are more cost-effective and faster than conventional applications (Willenbacher 2011). Since this work is a conceptual work, the further steps are the implementation of the ANN and the further development of the database of measures. In our ongoing work we will implement the ANN with Visual Studio, simulate it with the software MemBrain and afterwards structuring and implementing the database of measures. After the database of measures is implemented the next step is to integrate a Bayesian network. The Bayesian network is to be integrated to recognize possible consequences before they occur.

References

Atlan-tec (2017). http://www.atlan-tec.de/loesungen/prozessoptimierer/. Accessed 20 Feb 2017
Balog M, Gaunt AL, Brockschmidt M, Nowozin S, Tarlow D (2017) DeepCoder: learning to write programs. In: 5th international conference on learning representations 2017. https://openreview.net/pdf?id=ByldLrqlx. Accessed 25 Apr 2017
Brause R (1991) Neuronale Netze. Eine Einführung in die Neuroinformatik. Teubner, Stuttgart
DeepMind (2017). http://www.gruenderszene.de/allgemein/google-deepmind-energie. Accessed 19 Apr 2017
DeepMind2 (2017). https://www.googlewatchblog.de/2016/07/deepmind-kuenstliche-intelligenz-kuehlungskosten/. Accessed 20 Apr 2017
DeepMind3 (2017). https://deepmind.com/blog/deepmind-ai-reduces-google-data-centre-cooling-bill-40/. Accessed 22 Apr 2017
DeepMind4 (2017). http://www.theverge.com/2016/7/21/12246258/google-deepmind-ai-data-center-cooling. Accessed 22 Apr 2017
Grauel A (2008) Center computational intelligence and cognitive systems: Tutorial – Neuronale Netze, Fachhochschule Südwestfalen
Junker H, Meyer A, Sangmeister J (2015) Handbuch Standardsoftware im betrieblichen Umweltschutz. Erich Schmidt Verlag, Berlin, S. 111
Knowles J, Corne D (1999) The pareto archived evolution strategy: a new baseline algorithm for pareto multiobjective optimisation. In: Proceedings of the 1999 congress on evolutionary computation, vol 1, pp 98–105
Kruse R, Borgelt C, Braune C, Klawonn F, Moewes C, Stein-Brecher M (2015) Computational Intelligence – Eine methodische Einführung in Künstliche Neuronale Netze, Evolutionäre Algorithmen, Fuzzy-Systeme und Bayes-Netze, 2. Auflage. Springer Vieweg
Lippe W-M (2006) Soft-computing. Springer, Berlin [u.a.]
Mehnen J (2004) Mehrkriterielle Optimierverfahren für produktionstechnische Prozesse, Habilitationsschrift zur Erlangung der Lehrbefugnis für das Fach Informationssysteme in der Produktionstechnik an der Universität Dortmund
MemBrain (2017). http://www.membrain-nn.de/index.htm. Accessed 14 Mar 2017
Microsoft Visual Studio (2017). https://www.visualstudio.com/de/. Accessed 14 Mar 2017
Page B, Hilty LM (Hrsg.) (1995) Umweltinformatik. Informatikmethoden für Umweltschutz und Umwelt-forschung, 2. Auflage. Oldenbourg
Pollino C, Woodberry O, Nicholson A, Korb K, Hart B (2007) Parameterisation and evaluation of a Bayesian network for use in an ecological risk assessment. Environ Model Softw 22:1140–1152
Rey GD, Wender KF (2011) Neuronale Netze – Eine Einführung in die Grundlagen, Anwendungen und Datenauswertung, 2. Auflage. Hans Huber-Verlag

Rojas R (1993) Theorie der neuronalen Netze. Springer, Berlin

Russell SJ, Norvig P (2002) Artificial intelligence—a modern approach, 2. Auflage. Prentice Hall

TechFestival (2017). http://www.startup-energy-transition.com/. Accessed 19 Apr 2017

Tuma A, Haasis HD (1994) Real time production scheduling of environmental integrated production systems—a comparison of selected knowledge-based methods and machine learning algorithmus. In: Hilty LM, Jaeschke A, Page B, Schwabl A (Hrsg.) 8. Symposium Informatik für den Umweltschutz, Hamburg 1994. Band I und II. Metropolis, Marburg, S. 371–378

Tuma A, Haasis HD, Rentz O (1993) Entwicklung emissionsorientierter Produktionsabstimmungsmechanismen auf der Basis fuzzyfizierter Expertensysteme und Neuronaler Netze. In: Jaeschke A, Kämpke T, Page B, Radermacher FJ (eds) Informatik für den Umweltschutz. Informatik aktuell. Springer, Berlin, Heidelberg

Ultsch A, Beiersdorf S (1994) Optimierung von Ressourcen und wissensbasierte Spurenanalytik mit konnektionistischen Modellen und genetischen Algorithmen. In: Hilty LM, Jaeschke A, Page B, Schwabl A (Hrsg.) 8. Symposium Informatik für den Umweltschutz, Hamburg 1994. Band I und II. Metropolis, Marburg, S. 379–386

Ultsch A, Halmans G (1991) Data normalization with self-organizing feature maps. In: IJCNN-91-Seattle international joint conference on neural networks, vol 1

Umberto® (2017). https://www.ifu.com/umberto/?gclid=Cj0KEQjw2fLGBRDopP-vg7PLgvsBEi QAUOnIXEQuGS8g-_IpW6h2xgVSAWN-JPFfsMp5g6IbkWrxMuIaAj-v8P8HAQ. Accessed 14 Mar 2017

Watson (2017). http://mainichi.jp/english/articles/20161230/p2a/00m/0na/005000c. Accessed 21 Apr 2017

Werner H (1996) Modellierung ökologischer Daten mit Neuronalen Netzen. Auswertung von Wasser-Qualitäts-Messungen in Fließgewässern. In: Baumgart-Schmitt R (Hrsg.) Praktische Anwendungen Neuronaler Netze. Lit Verlag, Hamburg, S. 193 ff

Willenbacher M (2011) Prozessoptimierung mit künstlichen neuronalen Netzen. Akademiker Verlag, ISBN: 978-3639384727

Wohlgemuth V (2005) Komponentenbasierte Unterstützung von Methoden der Modellbildung und Simulation im Einsatzkontext des betrieblichen Umweltschutzes. Shaker Verlag

Wohlgemuth V (2014) Die Entwicklung der Stoffstromsimulation. In: Umweltinformatik - Einblick in drei Jahrzehnte der Entwicklung einer Wissenschaftsdisziplin. Shaker Verlag, April 2014, ISSN 1616-0886

Wohlgemuth V (2015) Ein Überblick über Einsatzbereiche von betrieblichen Umweltinformationssystemen (BUIS) in der Praxis. In: Lecture notes in informatics— proceedings, Gesellschaft für Informatik e.V., Bonn, ISBN 978-3-88579-640-4, ISSN 1617-5468

Zell A (1994) Simulation neuronaler Netze, München [u.a.]: 2., unveränd. Nachdruck der Ausgabe Bonn, Addison-Wesley 1994. Aufl. Oldenbourg, 2003

Zitzler E, Laumanns M, Thiele L (2001) SPEA2: Improving the Strength Pareto Evolutionary Algorithm, ETH Zürich, TIK-Report 103

Part V
Energy Aware Software-Engineering and Development

On the Impact of Code Obfuscation to Software Energy Consumption

Christian Bunse

Abstract The energy consumption of software systems is an area of increasing interest, especially for mobile application developers. A number of studies have been published that address possible optimizations of energy use and linked quality attributes such as performance. Of equal importance are, at least for commercial software systems, the protection of intellectual knowledge (IP) and the fight against software piracy (e.g., by code obfuscation to prevent reverse engineering). The mutual relations between energy consumption and IP protection force developers to strike a balance between them. This paper reports on the results of an empirical study on the effects of code level obfuscation by executing a number of usage scenarios across a set of ten Android applications. Results indicate that code-level obfuscation can have a significant impact on energy usage and performance and are likely to increase than decrease.

Keywords Energy efficiency · Software development · Obfuscation

1 Introduction

Energy is a limiting factor of information and communication technology. Especially mobile and embedded devices that do not have a permanent power supply, but depend on rechargeable batteries, are concerned. Due to increases in hardware performance and the incorporation of additional hardware components (e.g., Near Field Communication) the energy needs of devices are still growing. Software (i.e., operating system and application) utilizes hardware and, thus, is key in improving energy consumption.

Energy-aware software development, energy-aware algorithms and energy-aware resource substitution are only three examples of recently initiated research areas that try to reduce energy needs by optimizing the software rather than

C. Bunse (✉)
Hochschule Stralsund, Zur Schwedenschanze 15, 18435 Stralsund, Germany
e-mail: Christian.Bunse@fh-stralsund.de

© Springer International Publishing AG 2018
B. Otjacques et al. (eds.), *From Science to Society*, Progress in IS,
DOI 10.1007/978-3-319-65687-8_21

the hardware (Höpfner and Bunse 2011). Other research projects evaluate development guidelines to support developers in energy-aware software development (Bunse and Rohde 2016).

Software systems serve a (commercial) purpose. Protecting knowledge, impede illegal copying, etc. are therefore commonly addressed during post-development. Obfuscation is one means to do so. Vendors of obfuscation tools promise that source code will be reduced in size, while all quality attributes are kept unchanged. But is this really true? Obfuscation typically increases the complexity of a software system in order to reduce its understandability. It is unquestionable that the complexity of a systems is linked to performance and/or energy needs. Thus, an effect of obfuscation onto these can be assumed. In this paper, we report about the results of an empirical study that compares the energy consumption and performance of selected software systems being obfuscated or unchanged at the code level. The focus is on Java-based systems for Android devices. The goal is to evaluate or better to examine the balance between obfuscation and energy consumption.

In detail, the study is an *independent* replication of Sahin et al. (2016). As such the study aims at reproducing the findings by independent data. Therefore, we examine different applications and tools. The goal being to generalize existing findings to a wider population and therefore to enable meta-analysis. We assume that our study will support and extend the findings of Sahin et al.

In detail, we hypothesize that obfuscation has two faces regarding the energy consumption of a software system. One the one hand, source-code optimization such as dead-code removal and others will improve energy consumption. On the other hand, actions that increase complexity by changing the flow of control or by splitting and extending mathematical calculations, will have a counter effect and thus, increase energy and performance needs.

The remainder of this paper is structured as follows: Sect. 2 discusses related work. Section 3 presents background information regarding the examined quality factors as well as on measuring them. Section 4 introduces the underlying organization, principles and hypothesis of the study. Section 5 presents evaluation results and discusses threats to validity. Section 6 concludes the paper and provides a short outlook.

2 Related Work

Several research projects have been conducted regarding the energy consumption and-efficiency of (mobile) software systems. The main focus is the identification of "triggers" (code characteristics or development activities) that have an impact onto energy consumption. Studies on refactoring (Sahin and Pollack 2014), patterns (Bunse and Stiemer 2013), sorting algorithms (Höpfner et al. 2014), development guidelines (Bunse and Rohde 2016) and others show that there is not necessarily a link between performance and energy, and that design decisions have a significant impact onto energy consumption as well as onto the performance of a software system. Furthermore it appears that energy, performance, and cost are part of a "quality triangle" (i.e., the optimization of one factor may have an impact onto other factors as well).

Furthermore, there is a significant amount of research on optimizing energy usage at the level of programming languages and operating systems. Feeney (2001) proposes new energy-aware routing techniques, and Gurun et al. (2006) define energy-aware protocols for transmitting data in wireless networks. Another area is on a higher level of abstraction. Wilke et al. (2013) defines micro-benchmarks for emailing and web browsing applications and Bunse (2014) evaluates the impact of keeping users informed about their energy-related behavior.

Regarding post-production activities (Sahin et al. 2016) examined the effects of 18 code obfuscation techniques onto the energy consumption of mobile (Android) applications. The results of the study indicate that, while obfuscations can have a statistically significant impact on energy usage and are more likely to increase energy usage than to decrease energy usage, the magnitudes of such impacts are unlikely to be meaningful to mobile application users.

Generally, research has shown that there is a concrete link between software and the energy consumption of a device. Furthermore, studies have shown that developers should be aware of energy consumption and performance issues during development. However, more studies are needed to identify handles for optimization and estimation.

3 Background

3.1 Tools

A survey, using standard search engines, reveals that there are more than 100 tools currently available for code obfuscation. In the context of this study we used three publicly available tools (i.e., open-source). Reasons are their popularity, their openness (ease of measurement) and cost. These tools are: "ProGuard", "JODE", and "JavaGuard". Other (proprietary) tools such as "Allatori, Dash-O-Pro, GuardIT for Java, Stringer, or Zelix Klassmaster" were not used due to of cost/availability reasons.

ProGuard is a free Java class file shrinker and obfuscator. It can detect and remove unused classes, fields, methods, and attributes. It can rename the remaining classes, fields, and methods using short meaningless names. The resulting jars are smaller and harder to reverse-engineer.

JODE, the Java Optimize and Decompile Environment, is a java package containing a decompiler and an optimizer for java. The optimizer transforms class files in various ways with can be controlled by a script file.

JavaGuard is a general purpose obfuscator, designed to fit effortlessly into the regular build and testing process, providing peace of mind that Java code is more secure against de-compilation and other forms of reverse engineering.

3.2 Java Obfuscation Techniques

Software systems typically contain confidential information, which has to be protected. Obfuscation techniques are used to hide such information and thus, to prevent tampering or deter reverse engineering (i.e., security through obscurity or logic concealment). To achieve this goal one may use miscellaneous but concrete obfuscation techniques to disguise the original code. A good example of obfuscation might be found at: "http://www.semdesigns.com/Products/Obfuscators/JavaObfuscationExample.html". A detailed list of obfuscation techniques is provided by Batchelder (2006) which can be reduced to the following key activities:

- Renaming class, method, field and local names to shorter, obfuscated, or unique names or according to a given translation table.
- Removing debugging information.
- Removing dead code (classes, fields, methods) and constant fields.
- Optimizing local variable allocation.
- Changing the control flow of a system.

4 Study Design

4.1 Hypotheses

The goal of our study is to examine the impact of code obfuscation on the performance and energy consumption of software systems. The assumption is that obfuscation will have a visible effect regarding energy and performance. Thus we can state the null hypothesis as follows:

H_0 *There is no significant impact of obfuscation to the energy consumption of software.*

Based on H_0 we define the following alternative hypotheses:

H_1 *Obfuscation has a negative impact onto performance and energy consumption.*
H_2 *There are no significant differences regarding impact size and used obfuscation tools.*

These hypotheses are in line with research questions R1–R3 of the original study (Sahin et al. 2016). This, makes this study a true independent replication of the original one.

4.2 Materials

Systems. In the context of this study we examine if obfuscation at the code level has a significant impact onto the energy consumption and performance of a device (i.e., the device an application is running on). Comparison is done based on a "normal" or "vanilla" and one or more obfuscated versions of the system. Since this requires access to the source code, we used 10 systems that have been developed by students in the context of a course on mobile systems development (see Table 1).

Obfuscation was done by applying the selected tools (see Sect. 3.1) to get an obfuscated version which was then "bundled" to an Android APK-file. One of these tools (i.e. ProGuard) is also capable of shrinking the size of a file as well as optimizing source-code. These options might have an impact too, but were not evaluated in this phase of the experiment that examines the effects of obfuscation in isolation. In a second step, currently part of a master thesis, the effects of optimization and shrinking, as well as the combination of all available techniques are examined.

Usage Scenarios. For each system we specified an automated (i.e., executable) usage scenario based on Robotium technology (RobotiumTech 2016). The scenarios are driven by user input and were recorded during a study on the impact of user feedback on the energy consumption of software systems (Bunse 2014). They therefore represent real usage patterns. Application selection was based on factors such as availability, application type and accessibility. Measurement and obfuscation automation required slight modifications of the systems source-code.

4.3 Platform and Devices

Within this study we focus on mobile software systems. Therefore execution and measurement had to be based on a mobile device. As operating system we used Android due to its openness and the broad availability of Android ready devices. To

Table 1 Test systems

Application	Description	Size (Loc)	Size (MB)
AC-alarm	A simple alarm cock	2.500	0.3
MegaDownload	Downloader app for large files	1.600	0.2
My planner	Simple ToDo-list	1.214	0.2
Calculator	A calculator that supports the polish notation	3.900	0.4
Mensa-app	Lunch plan and shared meal rating	8.000	1.2
CatchMe	Jump and run game	12.400	2
NavigateMe	University indoor navigate	22.500	3
QIS-lens	Managing student records	15.600	2.5
Grade-tool	Managing class grades	11.450	2
LearningRabbit	Self-organization	7.900	1.2

cope with the wide variety of available systems we selected devices that support different API levels. In detail, we used: (1) a Google Nexus 7 (Android 4.1 at API-Level 16), (2) a Samsung Galaxy SII (Android 4.2.2 at API-Level 17), and (3) a Google Nexus 5 (Android 6.0.2 at API-Level 23). These devices also represent the two major devices classes, tablets and phones.

4.4 Measurement

In contrast to Sahin et al., we followed a software based measurement approach. According to Bakker (2014) and Schirmer and Höpfner (2012) differences between software and hardware based approaches are marginal and therefore negligible. In addition, a software based approach also allows to combine or integrate energy and performance measurement. Regarding energy consumption we used PowerTutor (Zhang et al. 2010), developed at the University of Michigan, which allows measurement at varying abstraction levels.

We use 'execution time' as basic performance metric. In the context of this paper we are using time-stamps based on global system time. Time-stamps are generated immediately before and after executing a test-application and are then stored in a central file. 'Execution time' is then calculated by subtraction.

Measuring properties like performance or energy consumption is affected by factors such as environmental conditions or system load. One means to mitigate such effects is replication, including the calculation of mean values. While replication extenuates external impacts, variation helps in generalizing results (i.e., basis for meta-analysis). Therefore we replicated each measurement more than 100 times (this also helps in weaken the impact of outliers due to environmental or operating system conditions) and performed measurements on different devices to mitigate hardware effects.

We checked that devices are in airplane mode, wireless communication and GPS are switched off, the battery is fully charged, the display is dimmed, and that the temperature of CPU and battery are "normal". During measurement, devices were not touched to avoid accelerometer reactions. Furthermore, we ensured that none of the in-built energy saving mechanisms were activated.

4.5 Runs

Overall, we conducted 4000 experimental runs (10 (Systems) * 4 (Variants: Three obfuscated plus one vanilla version) * 100 (Replications)). To ease analysis we used the overall consumption value that was calculated by PowerTutor and a mean value calculated by the performance measurement tool. However, this might be a bias since we cannot guarantee that these values are correct and comparable in the large.

5 Evaluation

5.1 Results

Table 2 provides an overview on the obtained measurement results of this study. In detail, the cells provide the mean energy (J) and performance (s) values measured for each system variant across all hardware platforms. The length of execution times (performance) depends on the complexity of the applied usage scenario and does not directly correspond to file sizes (see Table 1).

At a first glance the results follow those by Sahin et al. (2016). However, there are some results that warrant further examination. In general, the results show that improving execution time is often entailed by a reduction of energy consumption. It can be assumed that a reduced number of cycles (i.e., increasing performance) leads to a more efficient hardware utilization and thus to a lower energy consumption. But, regarding App 1 and *Jode* the energy consumption has increased while execution time was improved. App 1 (i.e. the simple alarm clock) makes use of services and "pending intents". Regarding *Jode* obfuscation led to performance gains of the main application but not of the service. It seems that a heavyweight service running in the background has a significant impact onto energy consumption. A comparable effect can be seen regarding App 6. App 6 is a game that heavily depends on UI interaction and graphics based on threading. We assume that concurrency and threading might be a weak spot.

In summary, data analysis reveals that the differences between obfuscated and non-obfuscated version are small. In detail, the mean deviation across all systems and variants is 0.119 J regarding energy consumption and 0.114 s regarding performance. Thus, in general there are visible, although quite small, effects visible.

On a more concrete level the data shows that in the worst case code obfuscation may lead to an increase of 15% regarding energy consumption and 20% in performance (app 9 and app 6). Thus, effects may become more prominent and this, in

Table 2 Averaged measurement results

	Non-obfuscated		Proguard		Jode		JavaGuard	
	Joule (J)	Time (s)	Joule (J)	Time (s)	Joule (J)	Time (s)	Joule (J)	Time (s)
App 1	33.12	44.325	31.99	42.08	37.11	41.84	36.07	39.01
App 2	60.14	72.123	61.00	74.24	60.95	78.93	66.45	85.23
App 3	44.25	60.765	42.98	61.89	62.34	68.43	57.98	72.98
App 4	54.57	74.002	53.78	77.78	54.89	80.81	94.67	76.65
App 5	88.50	120.003	90.61	135.51	98.67	132.01	90.02	140.11
App 6	184.37	250.412	189.39	245.99	182.99	277.34	179.23	290.34
App 7	221.25	300.967	223.98	300.46	228.62	302.89	250.66	299.66
App 8	129.06	175.432	128.76	170.19	130.95	169.77	130.00	190.99
App 9	147.50	200.324	148.90	199.92	148.09	210.99	151.31	209.67
App 10	106.93	145.275	107.99	150.87	110.03	152.94	110.45	155.32

turn, confirms the thesis that developers have to be aware of possible effects. Interestingly, the apps that strike attention both make use of Android services and depend on external communication (i.e., to third-party products). It seems that obfuscating (external) communication will deteriorate energy and performance requirements.

Another interesting effect is that there is no statistically significant correlation between code size and energy/performance needs. Thus, the code shrinking facilities of some obfuscators is neutral regarding resource requirements. Code shrinking is often done by removing dead code. Thus, only size but not complexity is changed.

Interestingly it appears that tools that perform code optimization (i.e., ProGuard) achieve "better" results than competitor tools and are, in some cases, also superior to the non-obfuscated original system. The optimization of ProGuard are closely related to development guidelines as proposed by Googler and others (e.g., prefer static, remove non-necessary try/catch blocks etc.). As shown Bunse and Rohde (2016) these guidelines, if systematically applied, lead to energy-efficient and performant systems.

5.2   Discussion

In general, the measurement results support hypotheses H_1 and H_2. As such we can confirm that in average non-obfuscated software systems are more energy-efficient and have a higher performance than obfuscated systems. But, by viewing the average effect across all variants/operating systems/devices etc. it appears that the effects are rather small, while worst case values are comparably large.

Interestingly, there is no direct relationship between the degree of obfuscation (LOC changed/removed/added) and the delta in energy consumption. This is supported by the finding that code shrinkage and the removal of dead code do not have a positive or negative impact onto energy consumption and performance. Thus, there must be another reason for the observed effects and worst-case values. One potential candidate is program complexity.

In detail, we used "McCabe Cyclomatic Complexity", "Afferent and Efferent Coupling", and Cohesion metrics to examine if obfuscation has an effect onto program complexity. Results show that there are only subtle visible effects. A reason might be that obfuscation does not really change the control-flow in order to preserve the original semantics. Thus, program complexity is likely not key in this regard. We can only assume that the reason might be in the very nature of single obfuscations:

- Packing local variables into bit fields,
- Adding dead-code switch statements,
- Indirecting if instructions, or
- GOTO instruction augmentation.

These basic obfuscations do not have a significant impact onto complexity as measured by the above mentioned metrics, but they increase access times, memory needs, and enlarge the number of steps to be executed. The combination of all these phenomenon is presumably responsible for the observed effects.

To understand if the differences between the results are statistical significant, we first checked the data for normality. It appears that the data is not normally distributed and thus, requires a nonparametric test such as the Wilcoxon-Matched-Pairs test. The test examines the difference in the mean (or median) of paired observations. In 76% (22 of 30) of the cases, when there is a statistically significant difference in energy usage, the obfuscated version is more likely to consume more energy than the original version. In 69% (20 out of 30) of the cases, when there is a statistically significant difference in performance, the obfuscated version is more likely have lower performance than the original version. Regarding hypothesis H_2 we also checked differences between obfuscation tools. Only in 10% (6 out of 60) of the cases significant differences were found. Thus, the impact of obfuscation onto energy and distribution appears to be equally distributed between tools.

In summary, we can state that the results of the analyses in terms of evidence (weakly) support the hypotheses H_1 and H_2. The null hypothesis can thus be rejected. Non-competitively to this study, we currently also investigate the effects of "code optimization" such as offered by ProGuard. Optimization comprises code shrinks, merges, and simplification. Preliminary results indicate that combining optimization and obfuscation mitigates effects as seen in the first run. In other words code optimization led in average to obfuscated systems that have a better performance without a higher energy demand.

5.3 Threats to Validity

This study is a replication of Sahin et al. (2016). This study aims (1) at confirming results, and (2) at broadening them as a basis for future meta-analysis. However, there are a number of threats that limit the generalization of results. In detail, the following threats have been identified:

The obfuscation tools used in this study (see Sect. 3.1) might not be representative for the application class. Other, commercially available tools, such as "Zelix KlassMaster" or "yGuard" might have led to other results. However, we believe that observable differences might occur regarding optimization. Regarding obfuscation all tools are comparable in their techniques (see Sects. 3.2 and 5.2) and philosophies. Thus, we assume that observable differences will be marginal.

Regarding performance, we decided to use execution time as metric. But, performance has more than one facet and results might be different when using other metrics. However, we already adapted the measurement approach used by Sahin et al. and still obtain comparable results. Thus, we believe that further variations will not lead to contra-dictionary results. Another problem might be that measurement creates overhead, which in turn may falsify measurement data. But, our results indicate that this overhead is negligible.

One threat might have occurred due to a biased selection/implementation of systems (study subjects). This might have resulted in a selection that is not representative of the population. The selected systems cover multiple domains and are different to those used by Sahin et al. Thus, we believe that the results, although not really sufficient for meta-analysis, help in broaden our knowledge about software related energy consumption.

Another threat might stem from the fact that tools internally apply different obfuscation techniques to a system that might be implemented in different ways. Thus, there might be an imbalance between the obfuscated variants of a system. This effect has been mitigated by using ten different systems.

Using pre-recorded usage scenarios might be the source for another group of threats. First, we used only one scenario per app. However, although the scenario indirectly utilizes hardware and is thus controlling consumption it does not have an effect onto the primary object under study (obfuscation).

To ease analysis we used the overall consumption value that was calculated by PowerTutor per run. This might be a bias since we cannot guarantee that these values are correct and comparable in the large.

In summary, the various threats to validity limit our ability to generalize the results in this instance, we plan to use the results of this experiment to facilitate further investigation. However, our data do support some plausible interpretations and can as such be used to guide developers.

6 Conclusion

In this paper we examined the effect of code-level obfuscation onto the energy consumption and performance of a software system. The results show that there are visible effects although these effects are small and possibly tolerable. As such our results support the stated hypotheses and also confirm the results of Sahin et al. (2016). This has brought research on code obfuscation and energy consumption closer (given the identified threats to validity) towards meta-analysis. However, the results may be used as a basis for identifying the best cost-effect ratio regarding security levels and resource needs.

In detail, we have set up an evaluation infrastructure and used it for a number of experimental runs. The infrastructure includes software-based energy and performance measurement tools, as well as a utility for automatically starting runs (based on Robotium technology). The infrastructure has been used to evaluate the effect of code-level obfuscation to energy needs and performance.

The results support our hypotheses: Code obfuscation is a valuable and powerful tool to protect intellectual property. However, it has to be carefully applied since obfuscation may lead to (small) increases regarding energy and time. We found that in general (across all applications and variants) energy consumption may be increased up to 15% and performance up to 20% (worst case). Thus, developers or companies who use obfuscation as a means for protection should be aware of possible negative effects.

In the context of our study we identified several open issues that warrant further research. First, we have to extend the study to examine obfuscation techniques in isolation. Second, the study has to be replicated across devices and operating systems to allow for meta-analysis. This is what we plan to do in the future.

References

Bakker A (2014) Comparing energy profilers for android. In Proceedings of the 21st twente student conference on IT, Enschede, The Netherlands

Batchelder M (2006) Java obfuscation techniques. http://www.sable.mcgill.ca/JBCO/examples.html

Bunse C (2014) On the impact of user feedback on energy consumption. In: 28th international conference on informatics for environmental protection, Oldenburg, Germany

Bunse C, Rohde A (2016) Software development guidelines for performance and energy: initial case studies. In: EnviroInfo2016, 30th international conference, Berlin

Bunse C, Stiemer S (2013) On the energy consumption of design patterns. Softwaretechnik-Trends 33(2)

Bunse C, Höpfner H, Roychoudhury S, Mansour E (2014) Choosing the best sorting algorithm for optimal energy consumption. In: Proceedings of the 4th international conference on software and data technologies

Feeney L (2001) An energy consumption model for performance analysis of routing protocols for mobile ad hoc networks. Mobile Netw Appl 6(3)

Gurun S, Nagpurkar P, Zhao B (2006) Energy consumption and conservation in mobile peer-to-peer systems. In: 1st international workshop on decentralized resource sharing in mobile computing and networking

Höpfner H, Bunse C (2011) Energy awareness needs a rethinking in software development. In: 6th international conference on software and data technologies, Seville, Spain

RobotiumTech (2016) Robotium—user scenario testing for android. https://github.com/RobotiumTech/robotium

Sahin C, Wan M, Tornquist P, McKenna R, Pearson Z, Halfond WGJ, Clause J (2016) How does code obfuscation impact energy usage? J Softw: Evol Process 28(7):565–588

Sahin C, Pollock L (2014) How do code refactorings affect energy usage? In: Proceedings of the 8th international symposium on empirical software engineering and measurement

Schirmer M, Höpfner H (2012) Software-based energy requirement measurement for smartphones. In: 42nd GI Jahrestagung

Wilke C, Richly S, Götz S, Assmann U (2013) Energy profiling as a service. In: 43rd GI Jahrestagung 2013, Koblenz, Germany

Zhang L, Tiwana B, Qian Z, Wang Z, Dick R, Mao Z, Yang L (2010) Accurate online power estimation and automatic battery behavior based power model generation for smartphones. In: Proceedings of the 8th IEEE/ACM/IFIP international conference on hardware/software codesign and system synthesis. ACM

Energy Consumption and Hardware Utilization of Standard Software: Methods and Measurements for Software Sustainability

Achim Guldner, Marcel Garling, Marlies Morgen, Stefan Naumann, Eva Kern and Lorenz M. Hilty

Abstract The ubiquity of information and communication technologies (ICT) results in substantial amounts of energy consumption and thus, CO_2-emissions. Since software induces the energy consumption of hardware, some reliable procedures and tests for measuring software are necessary. We present such a method and prove our measurement concept by applying it to two software product groups: word processors and content management systems. Even though the two groups are very different in terms of their requirements, we were successful in the creation of a measurement environment that supports the production of reliable, verifiable results, allowing the comparison of the energy consumption induced by software

A. Guldner (✉) · M. Garling · M. Morgen · S. Naumann · E. Kern
University of Applied Sciences Trier, Environmental Campus Birkenfeld,
Institute for Software Systems, Birkenfeld, Germany
e-mail: a.guldner@umwelt-campus.de

M. Garling
e-mail: marcel.garling@posteo.de

M. Morgen
e-mail: marlies.morgen@gmx.de

S. Naumann
e-mail: s.naumann@umwelt-campus.de

E. Kern
e-mail: mail@nachhaltige-medien.de

E. Kern
Leuphana University Lüneburg, Lüneburg, Germany

L.M. Hilty
Department of Informatics, University of Zurich, Zurich, Switzerland
e-mail: hilty@ifi.uzh.ch

L.M. Hilty
Technology and Society Lab, Empa Swiss Federal Laboratories for Materials Science
and Technology, St. Gallen, Switzerland

L.M. Hilty
KTH Royal Institute of Technology, Stockholm, Sweden

© Springer International Publishing AG 2018
B. Otjacques et al. (eds.), *From Science to Society*, Progress in IS,
DOI 10.1007/978-3-319-65687-8_22

systems with similar functionality. The method shows viable results for desktop and client-server systems, paving the way for further setups like e.g. mobile and embedded devices.

Keywords Green software engineering · Software energy consumption · Green IT

1   Introduction

The ubiquity of information and communication technologies (ICT) results in substantial amounts of energy consumption and thus, CO_2-emissions. According to Hintemann and Clausen (2016), who give an overview analyzing different international studies addressing the topic, "IT energy consumption [...] accounts for approx. 8% of global electricity consumption". Since software induces the energy consumption of hardware, and following the rule "what you can't measure you can't manage" to answer the question how this energy consumption of ICT can be measured, it is necessary to define and test reliable methods and procedures. Current activities, such as the project UFOPLAN-SSD,[1] develop criteria and indicators regarding the resource and energy efficiency of software. To understand especially the influences of hardware utilization and energy consumption, it is necessary to measure with realistic usage scenarios. Therefore, we take a closer look at CPU and RAM utilization, and make concrete measurements of energy consumption, following criteria developed during UFOPLAN-SSD. We compare two products with similar functionality regarding energy consumption from two product groups (word processing and content management systems).

The main objective of the project is to come up with criteria to describe sustainable software in a more specific way. Up to now, there exist many different definitions, approaches for criteria, and thus understandings what sustainable software is. Here, we want to provide a criteria collection to better define our understanding of impacts of software onto sustainable development. Different targets groups are in mind: (a) recommendations how to develop sustainable software for programmers, (b) an eco-label for software to create transparency especially for end users, and (c) a basis to come up with green software procurement guidelines for purchaser.

The following paper will present testing methods for proposed criteria, being results of the UFOPLAN-SSD project, to evaluate the collection.

---

[1]See http://green-software-engineering.de/en/projekte/ufoplan-ssd-2015.html [2017-04-23].

2 Related Work

In order to be able to assess the sustainability of software products (Naumann et al. 2011; Hilty et al. 2015; Calero and Piattini 2015), corresponding measurement methods are required. In the past years, research presented many approaches how to measure energy consumption of software (Lami and Buglione 2012; Corral et al. 2013; Willnecker et al. 2014). Calero et al. (2013) addressed the measures of sustainability of software products by implementing a structured literature review. One of their findings is that an important topic in this context is the measurement of power consumption. Kern et al. (2013) described and compared different approaches of measuring energy consumption of ICT. Recently, Feitosa et al. (2017) and Procaccianti et al. (2016) presented a comprehensive study on power measurements. Generally, the research approaches can be differentiated by measurement contexts and objects (What has been measured?), motivation (Why is the measurement implemented?), and methods (How have the measurements been implemented?). Table 1 lists recent papers that present different software energy measurements.

The examples, being just small extracts from the current literature regarding software measurements, show that there are quite different motivations, approaches, and outcomes in measuring software products. This points out that the addressed field is large and diverse. Thus, there are many possibilities to optimize software products. Overall, the method presented by Williams and Tang (2015) is most comparable to the approach that we present in this paper.

In addition to exemplary case studies, research results in measurement models have been published. Most of the models can be used to estimate the energy

Table 1 Current literature addressing software energy measurements

Reference	Objects	Motivation	Findings
Rashid et al. (2015)	Sorting algorithms implementations	Analyzing the energy efficiency in theory (algorithm) and in practice (implementation)	The algorithm and the language affect the total energy consumption
Williams and Tang (2015)	Operating systems (OS)	Creating a methodology to measure and describe OS's impact on device energy consumption and related environmental impact	OS and/or activity can significantly impact the energy consumption of a device
Awad et al. (2017)	Software-defined networking routing application	Developing a solution to minimize network energy consumption while satisfying traffic requirements	Solution to handle single and multiple data flow routing requests and achieves high power savings
Feitosa et al. (2017)	Gang of four patterns	Analyzing an indicator that has attracted the attention of researchers and practitioners	Majority of cases: alternative design (without using patterns) has lower energy consumption

consumption or similar impacts of software applications (Beghoura et al. 2017; Chowdhury and Hindle 2016). On the other hand, models are used to describe the measurement method itself (Dick et al. 2011; Godboley et al. 2017). Caused by the complexity of ICT systems, it is not always possible to measure every part of these systems. Here, estimation models gain a great relevance. One example is data traffic in systems using online resources.

While the awareness for energy and further environmental issues of desktop software is quite small, the issue gains much more attention in case of mobile applications (Manotas et al. 2016). Consequently, energy measurements seem to be an interesting research question in this context (Gui et al. 2016; Li et al. 2016; Siebra et al. 2012).

Overall, comprising desktop, server, and mobile applications, as proposed within the GREENSOFT Model (Naumann et al. 2011), the aim is to produce tools (e.g. as proposed in Noureddine et al. (2016)) and recommendations (e.g. as presented in Procaccianti et al. (2016)) to support developers in creating and maintaining green software products.

3 Measurement and Analysis Method

As mentioned above, we will focus on the measurement of hardware utilization and introduce a measurement method for the electricity consumption. To ensure the comparability between the software products from one software product category (i.e. "word processors" (WP) or "content management systems" (CMS)), we defined a standard usage scenario for each category.

There are several challenges involved in the creation of such a scenario. First, it is important to analyze the tasks which users usually carry out with the software and consider the software's functions and modules that are frequently used. In addition, we want to discover optimization potentials within the software products. Therefore, we look out for functionalities that may cause high energy consumption or hardware utilization. Finally, we want to achieve a scenario that is as realistic as possible, considering user behavior. Therefore, we record users performing the scenario where possible, and where necessary, e.g. for CMS, we use other means like a load generator tool that randomly calls the websites generated by the CMS many times per minute, simulating many users. In Sect. 4, we present two exemplary standard usage scenarios.

We carried out the actual measurements of the energy consumption and hardware utilization on a test rig which we devised following ISO/IEC 14756 as introduced in Dirlewanger (2006). In it, the software product is installed on a system under test (SUT), which consists of the hardware (desktop computer, server, smartphone, etc.), operating system and, if necessary, further runtime environments. The SUT not only runs the software product, but also records its hardware utilization during the measurements. Additionally, we use an industrial wattmeter[2] with a sampling rate of 20 kHz per phase and resolutions of 0.01 V and 1 mA, respectively, to measure the energy consumption of the SUT. The collected data is averaged per second and then, stored and analyzed in a central data aggregator and evaluator. A detailed description of the measurement process can be found in Dick et al. (2011).

Another factor to consider is the overhead on the SUT, resulting from the recording of the hardware utilization. In our measurements, we found the overhead of the data recording tools (performance monitor on Windows and collect on Ubuntu) negligible. However, if more precise measurements are necessary, it may be advisable to measure the recording software first and then subtract this baseline from the actual measurement.

When the setup is complete, we use a load generator to execute the scenarios automatically and repeatedly on the SUT, to ensure statistically robust results (cf. Sect. 4). In the following, we describe the method for analyzing the measurement results to evaluate the average energy consumption and the average hardware utilization of the software products while performing the standard usage scenario.

Energy consumption is calculated as the arithmetic mean value over each measurement sample. The energy for each measurement sample is calculated as the sum over the per second average wattage (P_i) that was measured for the SUT. We then divide the result by 3600 s/h to receive the result for the indicator in watt-hours (Wh). Hence, the formula for calculating the indicator for energy consumption is

$$E = \frac{1}{3600 \cdot n} \sum_{k=1}^{n} \sum_{i=1}^{m} P_i,$$

where the duration of the k-th measurement is m seconds and there are a total of n measurements.

Processor utilization is calculated as the arithmetic mean value over each measurement sample in per cent (%) of the maximum processor performance of the SUT. This means that if there are a total of n measurements, we calculate the indicator for the processor utilization as

$$U_P = \frac{1}{n} \sum_{k=1}^{n} \overline{x_k},$$

where $\overline{x_k}$ is the arithmetic mean of the k-th sample in %. We calculate the indicators for the other hardware capacities like **RAM utilization** in an analogous manner.

---

[2]Janitza UMG 604 (https://www.janitza.com/umg-604.html [2017-04-23]).

The amount of **data transferred** over the network and read from/written to the hard disk is recorded by the SUT in megabytes per second (MB/s) and the corresponding indicator is calculated in analogy to the hardware capacities.[3]

4 Exemplary Results

In the following, we describe exemplary results of our measurements, including the structures of the corresponding usage scenarios. As mentioned above, we chose two popular software products each from the product groups WP and CMS to serve as our examples. These two software groups are very distinct and therefore impose different requirements to the usage scenarios, the measured data and the measurement setup (client-server environment in case of the CMS). We selected the most current, recommended version at the time for each software.

4.1 Word Processor Group

For the group of word processors, we selected two widespread word processing software products for comparison, abbreviated with WP1 and WP2. The decision to use these products has been made caused by the fact that they are widely used and one of them is a commercial software product whereas the other one is open source. The following paragraph presents the recorded measurement results for both word processors.

As already mentioned above, we created a **usage scenario** including actions that are common when using this kind of software and, to enable the comparison, are feasible with all products of this group. With that in mind, we decided to implement actions like editing the content and the layout of the document as well as saving it several times during the scenario. As basis for the scenario, the document "Surgical Anatomy" by Joseph Maclise[4] was used. It contains more than three hundred pages including text, pictures, tables, etc. Eventually, we created two mostly identical PDF documents, one with each word processor.

To get representative results, the two products were installed sequentially on the SUT and started in standard configuration. Additionally, we paid attention to a similar view of the application which comprises e.g. opening the software in full screen, adjusting the zoom level and activating the control characters.

---

[3]We use the performance monitor included in MS Windows (https://technet.microsoft.com/en-us/library/cc749249(v=ws.11).aspx [2017-04-26]) or the "collectl" package for Linux (http://collectl.sourceforge.net/ [2017-04-26]).

[4]Surgical Anatomy by Joseph Maclise (http://www.gutenberg.org/ebooks/24440 [2017-04-22]).

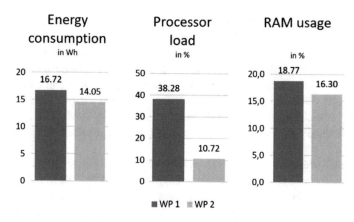

Fig. 1 Measurement results of word processors WP1 and WP2

When comparing WP1 and WP2, we found that there were significant differences between the results in energy consumption, processor load, and RAM usage (see Fig. 1).

A closer look at the energy consumption **results** during the usage scenario points out significant differences between the products. Both applications had a high peak in the beginning while starting the software, opening the document and changing the whole layout of the content. Besides that, measurements of WP2 show the highest peaks while saving the document. Other actions have only small influence on the general level of energy consumption. However, WP1 exhibited much greater swings while performing actions of all kinds. Even without any user input the flashing cursor does keep the value of consumption at an elevated level. That is part of the reason why we also analyzed the transition to sleep mode for both software products. For this scenario, we did open a new document without any content and had a look at the software reaction. WP1 waits for 26 s until the cursor stops flashing. Afterwards, energy consumption and hardware utilization decreases to the level of the idle reading. WP2 stays in active mode for 18 s without user input before it switches to the sleep mode and the hardware utilization and energy consumption decrease to a minimum.

The results indicate that there is optimization potential for WP1. Especially for the flashing cursor. In addition, in WP2 most of the editing, like adding content or changing the layout of the document, had a lower impact on the energy consumption than it did in WP1.

4.2 Content Management Systems Group

For our software comparison, we chose two CMS (CMS1, CMS2) owing to their large market share. To be able to perform the necessary measurements with the

CMS, we had to create content in the form of a website to be worked with. To serve the comparability of the results, this content has to be equal for all products. The demand for equality covers the content and the structure as well as the design of the website. We fulfilled these requirements by developing a basic template, implementing it for each CMS and then compiling the same structure and content in each one.[5] In terms of size, the resulting website represents a small information site, comprised of text and images, like those often operated by small businesses. Except for a commentary function, the website is rather simple without any further user interaction.

The **usage scenario** based on the website covers the typical user interactions of an editor working in the administrator panel (often called "backend"). After the user logs in, some comments are managed, then a new page is created, edited, and published. The user then uploads PDF files (7×270 KB) and links them. Afterwards, they are removed again with the built-in revision tool. Finally, all changes are revoked, allowing the next measurement run to start with identical conditions.

We created this scenario to run on a client, which serves as the load generator, while the CMS were installed on the server (the SUT). We then measured both systems simultaneously during the test runs. The procedure was applied repeatedly to measure the impact of different levels of stress: In the first setup, only the automated user from the usage scenario executes interactions with the system. In a second setup however, we used the aforementioned load generator tool to automatically perform 14,000 additional website requests per minute, simulating several client computers.

The following paragraphs cover the **results**, starting with those measured only for one user. Here, it is remarkable that executing the scenario using CMS1, the server sends, on average, substantially more data than it does running CMS2 (CMS1: 0.09 MB/s; CMS2: 0.03 MB/s).

Although the network activity on the client side is higher for CMS1, it is CMS2 that shows the most hard drive writing activity (CMS1: 0.08 MB/s; CMS2: 0.13 MB/s).

We soon noticed the fact that one single user does not provide enough activity to noticeably effect the hardware utilization and the resulting energy consumption of the server. The only observation worth mentioning is a peak in the energy consumption during the PDF file upload.

The results of the second setup (the one with the load generator tool in use) offer a much more differentiated insight. The energy consumption and the hardware utilization are now clearly elevated. However, conclusions to any of the specific activities in the usage scenario and their impact cannot be drawn anymore.

Comparing the two systems under stress, CMS2 turned out to be more demanding than CMS1 in many aspects. It has a higher energy consumption,

---

[5]Technically, the templates are based on the prefabricated designs from the "Bootstrap" web framework (http://getbootstrap.com/ [2017-04-22]).

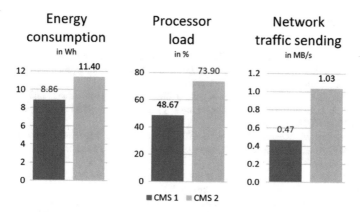

Fig. 2 Measurement results of CMS1 and CMS2

a higher CPU usage and now also a higher average network activity in sending direction from server to client (see Fig. 2).

5 Conclusion and Outlook

We presented a method for assessing the sustainability of software products, based on measuring the energy consumption and hardware utilization which is induced by a typical usage scenario of a software product. We proved our measurement concept by applying it to two software product groups, word processors and content management systems. Even though the two groups are very different in terms of their requirements, we were successful in the creation of a measurement environment that supports the production of reliable, verifiable results for each group and the possibility to compare the sustainability of software products with similar functionality within the groups. The method shows viable results for desktop and client-server systems, paving the way for further setups like e.g. mobile and embedded devices. Based on these results, developers can improve their code, and software purchasers can make decisions, also based on the sustainability of the products.

In future investigations, the energy consumption induces by network and remote systems, which also contribute to the overall energy balance, has to be considered in more detail.

Acknowledgements The work for this paper was supported by the UFOPLAN-SSD project, funded by the German Federal Environmental Agency (Umweltbundesamt - UBA).

References

Awad MK, Rafique Y, M'Hallah RA (2017) Energy-aware routing for software-defined networks with discrete link rates: a benders decomposition-based heuristic approach. Sustain Comput: Inf Syst 13:31–41. doi:10.1016/j.suscom.2016.11.003

Beghoura MA, Boubetra A, Boukerram A (2017) Green software requirements and measurement: random decision forests-based software energy consumption profiling. Requir Eng 1–14. doi:10.1007/s00766-015-0234-2

Calero C, Bertoa M, Angeles Moraga M (2013) A systematic literature review for software sustainability measures. In: 2013 2nd international workshop on green and sustainable software (GREENS), pp 46–53

Calero C, Piattini M (2015) Introduction to green in software engineering. In: Calero C, Piattini M (eds) Green in software engineering. Springer, pp 3–27

Chowdhury SA, Hindle A (2016) GreenOracle: estimating software energy consumption with energy measurement corpora. In: Proceedings of the 13th international conference on mining software repositories, pp 49–60

Corral L, Georgiev AB, Sillitti A, Succi G (2013) A method for characterizing energy consumption in Android smartphones. In: 2013 2nd international workshop on green and sustainable software (GREENS), pp 38–45

Dick M, Kern E, Drangmeister J, Naumann S, Johann T (2011) Measurement and rating of software-induced energy consumption of desktop pcs and servers. In: Pillmann W, Schade S, Smits P (eds) Innovations in sharing environmental observations and information: proceedings of the 25th international conference enviroinfo October 5–7, 2011, Ispra, Italy. Shaker, Aachen, pp 290–299

Dirlewanger W (2006) Measurement and rating of computer systems performance and of software efficiency: an introduction to the ISO/IEC 14756 method and a guide to its application. Kassel University Press, Kassel

Feitosa D, Alders R, Ampatzoglou A, Avgeriou P, Nakagawa EY (2017) Investigating the effect of design patterns on energy consumption. J Softw: Evol Process 29(2)

Godboley S, Dutta A, Mohapatra DP, Mall R (2017) Green-J^3 model: a novel approach to measure energy consumption of modified condition/decision coverage using concolic testing. CSI Trans ICT 1–17

Gui J, Li D, Wan M, Halfond WGJ (2016) Lightweight measurement and estimation of mobile ad energy consumption. In: Proceedings of the 5th international workshop on green and sustainable software, pp 1–7

Hilty L, Lohmann W, Behrendt S, Evers-Wölk M, Fichter K, Hintemann R (2015) Green software: final report of the project: establishing and exploiting potentials for environmental protection in information and communication technology (Green IT). Report commissioned by the Federal Environment Agency, Berlin, Förderkennzeichen 3710 95 302/3(23)

Hintemann R, Clausen J (2016) Green cloud? The current and future development of energy consumption by data centers, networks and end-user devices. In: Grosso P, Lago P, Osseyran A (eds) Proceedings of ICT for sustainability 2016. Atlantis Press

Kern E, Dick M, Naumann S, Guldner A, Johann T (2013) Green software and green software engineering—definitions, measurements, and quality aspects. In: Hilty LM, Aebischer B, Andersson G, Lohmann W (eds) ICT4S ICT for sustainability: proceedings of the first international conference on information and communication technologies for sustainability, ETH Zurich, February 14–16, 2013. ETH Zurich, University of Zurich and Empa, Swiss Federal Laboratories for Materials Science and Technology, Zürich, pp 87–94

Lami G, Buglione L (2012) Measuring software sustainability from a process-centric perspective. In: 2012 joint conference of the 22nd international workshop on software measurement and the 2012 seventh international conference on software process and product measurement (IWSM-MENSURA), pp 53–59

Li D, Lyu Y, Gui J, Halfond WGJ (2016) Automated energy optimization of http requests for mobile applications. In: Proceedings of the 38th international conference on software engineering, pp 249–260

Manotas I, Bird C, Zhang R, Shepherd D, Jaspan C, Sadowski C, Pollock L, Clause J (2016) An empirical study of practitioners' perspectives on green software engineering. In: Proceedings of the 38th international conference on software engineering, pp 237–248

Naumann S, Dick M, Kern E, Johann T (2011) The GREENSOFT model: a reference model for green and sustainable software and its engineering. SUSCOM 1(4):294–304. doi:10.1016/j.suscom.2011.06.004

Noureddine A, Islam S, Bashroush R (2016) Jolinar: analysing the energy footprint of software applications. In: The international symposium on software testing and analysis

Procaccianti G, Fernández H, Lago P (2016) Empirical evaluation of two best practices for energy-efficient software development. J Syst Softw 117:185–198

Rashid M, Ardito L, Torchiano M (2015) Energy consumption analysis of algorithms implementations. In: 2015 ACM/IEEE international symposium on empirical software engineering and measurement (ESEM), pp 1–4

Siebra C, Costa P, Miranda R, Silva FQB, Santos A (2012) The software perspective for energy-efficient mobile applications development. In: Proceedings of the 10th international conference on advances in mobile computing and multimedia, pp 143–150

Williams DR, Tang Y (2015) A methodology to measure the environmental impact of ICT operating systems across different device platforms. J Electr Sci Technol 3

Willnecker F, Brunnert A, Krcmar H (2014) Model-based energy consumption prediction for mobile applications. In: Proceedings of workshop on energy aware software development (EASED)@ EnviroInfo, pp 747–752

Green Computing, Green Software, and Its Characteristics: Awareness, Rating, Challenges

Eva Kern

Abstract Issues of green computing, especially on the hardware side, arouse interest of scientific communities since several years. The number of scientific publications and research activities rises. This begs the question if resulting ideas, approaches, and findings find their way from science to society. Hence, similar to different researchers, we conducted a survey regarding corresponding issues. Since we focused on software users, we will compare our results to those addressing students, practitioners, and IT managers. The following paper will provide an insight into awareness of green software and green computing, present a user survey we conducted in the summer of 2016 in Germany, and approaches on how to create awareness to environmental issues of ICT, especially on the software side.

Keywords Green computing · Eco-labelling · Awareness · Green software

1 Introduction

Information and communication technologies (ICT) involve with a negative and a positive side: On the one hand, they have a significant impact on the worldwide energy consumption and global CO_2 emissions (Andrae and Edler 2015; Malmodin et al. 2013). On the other hand, according to the #SMARTer2030 report (2015), "ICT can enable a 20% reduction of global CO_2 emissions by 2030, holding emissions at 2015 levels". Thus, it is "both a solution and a problem for environmental sustainability" (Selyamani and Ahmad 2015). In order to strengthen the positive role of ICT, the Global e-Sustainability Initiative sees especially three important stakeholders: policymakers, business leaders, and consumers. GeSI,

E. Kern (✉)
Leuphana University Lueneburg, Scharnhorststrasse 1, 21335 Lüneburg, Germany
e-mail: mail@nachhaltige-medien.de

E. Kern
Institute for Soft-ware Systems, Environmental Campus Birkenfeld,
University of Applied Sciences Trier, Birkenfeld, Germany

Global e-Sustainability Initiative (2015) Above, big players in software development, like Microsoft and SAP, seem to be important to be included.

Overall, environmental issues of software comprise not only energy issues but also
aspects of resource consumption (including e.g. hardware needed to operate a software
product, adaptability of the hardware in use, and the way of delivery of the software
product), issues regarding the potential hardware operating life (like backward compatibility and platform independency), and characteristics of the software supporting
user autonomy to enable a more environmental friendly way of using the products. In
Kern et al. (2015) we identified possible criteria for green software.

We see the creation of awareness for environmental issues of ICT and especially
software as the first step in moving towards a more environmental friendly way of using
ICT. We agree with Ramachandiran (2012) in regards of awareness: "Awareness is a
need to practice an idea of the new things. Without awareness, the practice does not do
appropriately and adequately and the other way around". In the context of green
computing, Lago and Jansen (2010) differentiate between three kinds of awareness:
"the direct environmental impact of executing software services (service awareness);
the indirect environmental impact of their development process (process awareness);
and the use of services to monitor and measure the two above, so to create people
awareness)". The following paper comprises all of them when talking about awareness.
The focus is set on the view of end users (being consumers) and practitioners (being
developers, managers etc.) to show possibilities "from science to society".

Thus, the following paper addresses the two questions: (i) *Is there an awareness
towards green computing and especially green software?* (ii) *How can the
awareness towards these topics be raised and/or increased?*

2   Awareness on Green Computing and Green Software

While there are more and more scientific conferences, workshops, and journals
addressing Green IT topics and integrate increasingly also green software aspects,
the issues still gain little to no awareness by non-academics. Recent research
activities "mainly focused on solutions for the IT industry and business, and
neglected the end users" (Selyamani and Ahmad 2015). The surveys we present in
the following section prove these impressions and theses.

Talking about "Green Computing" or rather "Green ICT" mainly sets the focus
on hardware issues while software issues are directly named as "green software" or
similar terms. Here, both aspects are included: the specific focus of the study in
question is mentioned in the second column of the table (*italicized*).

Different surveys address the knowledge and awareness of green computing.
They aim at target groups like programmers, practitioners, ICT managers, and
students (Table 1). Especially the last ones were questioned in different projects.
Here, the aim is to improve education and information activities in the context of
Green IT (Selyamani and Ahmad 2015; Dookhitram et al. 2012). After finishing
their studies, the former students might bring their knowledge about green

Table 1 Overview of studies addressing the awareness of green computing issues

Target group	Objective and focus	Findings	Source
Programmers	Do programmers know about energy consumption of *software* and how to reduce it although such issues are not addressed in education or trainings?	Programmers had limited knowledge of energy efficiency, lacked knowledge of the best practices to reduce energy consumption and were unsure about how software consumes energy	Pang et al. (2016)
Practitioners	Relationship between research community and practitioners: same interests? Asking for the practitioners' perspective to get a motivation and guide for green *software* engineering research	Practitioners care about energy but are not as effective as they could be at creating efficient applications	Manotas et al. (2016)
Students in Mauritius	Analyze the awareness of green computing (focus: *hardware*) among students	Students have a moderate knowledge about green computing but their everyday green computing practices are not satisfactory	Dookhitram et al. (2012)
Students in Malaysia	*Hardware* focus: usage and selling of computers and related resources	Students have an average level awareness about green computing and their everyday practices are not satisfactory	Selyamani and Ahmad (2015)
Buyer and supplier	Analyze people's awareness of *software* engineering issues	People are aware of the importance of energy efficient software, but lack information and resources to engage this topic	Ferreira (2011)
ICT Managers	How do ICT managers implement energy efficient methods (focusing on *hardware* issues) in their organizations?	Just over a third of organisations in Europe do not implement green IT practices	Kogelman (2011)

computing into companies and organizations. Here, ICT managers are in charge of implementing green strategies. Therefore, Kogelman (2011) surveyed ICT managers. For Manotas et al. (2016), the motivation for implementing an awareness study among practitioners in general was to gain insights for researching activities regarding green software engineering.

Summarizing the findings of the considered studies, the knowledge about green computing of the interviewed programmers, students, practitioners, buyers, suppliers, and managers are a great deal. Most of them seem to have some knowledge. However, the lack of practice is large. The studies addressing students mainly focus on the hardware aspects and integrate software aspects just slightly (e.g. operating system and search engines Selyamani and Ahmad 2015) or concentrate on specific single concerns of environmental issues of software (e.g. energy Kogelman 2011;

Ferreira 2011). Nevertheless, all of them provide an impression of the awareness of green computing and green software. Like Manotas et al. (2016) describe it, the studies "can motivate and guide green software engineering research".

Indeed, a study addressing the group of software users was, as far as we know, missing. Thus, we conducted a survey on social interests in green software and acceptance factors influencing an eco-label for software products. The study as well as its result is presented in the following section. In addition, we combine our findings with the results of the surveys listed in Table 1.

3   User Survey: Environmental Issues of Software

From August, 16 to October, 5 2016 we conducted a user survey addressing environmental issues of software and their certification. The overall aim was to bridge from science to society. This requires the knowledge about society's interests. Thus, the data was collected via an anonymous online questionnaire, addressing German software users by asking the questions in German.

3.1   Objectives and Methodology

The objective of the survey was to gather the aspects of green software that arouse social interest and factors that influence the acceptance of a possible certification of green software (Kern 2016). Thus, the first step is to find out if there is an awareness for environmental issues of software. In contrast to the other surveys, here, the focus target group were end users of software.

The questionnaire was divided into three sections: (1) factors of influence on the acceptance of a certification of green software products, (2) evaluation of possible criteria for an eco-label for software products, and (3) demographic characteristics. A total of 712 questionnaires[1] were completed and used for data analysis. The data was analyzed using both, descriptive methods and correlations between variables.

In the following, the results of Sections 2 and 3 are described. We start with the demographic data to first give an impression of the survey responders. The outcomes of Section 1 are not presented in detail, since the paper focuses on the awareness of and interest in environmental issues of software instead of labeling them.

---

[1]The sample comprised n = 854 questionnaires. This dataset has been revised by excluding those that were not completed.

3.2 Results

Asked for their role in using software, most of the survey respondents (78.2%) see themselves primarily as software users (possible answers: developer, user, purchaser, distributor, administrator, vendor, IT manager, and others). 55.8% of them were female, 40.7% male, and 3.5% did not answer this question. More than two-thirds are between 20 and 39 years old (39.3% are 20–29 years old, 29.4% between 30 and 39 years). About half of them have a higher university degree (48.3%); 52.5% of the survey participants work in the context of natural sciences, geography and computer sciences.

Most of the respondents know the Blue Angel[2] (94.0%), whereas the other eco-labels that were part of the survey are known by 68.0% in case of the ENERGY STAR[3] and 10.3% in case of the TCO Certification.[4] Only 4.6% of the respondents did not know any of the three labels. However, there is quite a lack of knowledge regarding the details of the awarding criteria and awarded products. This could be interpreted as demand to better inform about these aspects and thus about environmental issues. Indeed, according to their answers, the survey participants have a high environmental awareness: 61.4% of them are willing to pay more for organic products, 23.7% catch up on environmental topics.

In spite of the environmental interests and knowledge about corresponding labels, more than half of the participants of the survey never paid heed to an eco-label for software products (56.3% "agree", 18,7% "somehow agree", 12,6% "somehow do not agree", 10.4% "do not agree" to "I never cared if there is a certification of environmental issues of software."). If there are two products and the costumers know just the one that is not labeled as "green", they would buy the product that is not-green but known instead of the green software product (63.3% agree). However, 23.2% of the respondents can imagine taking "certification of environmental issues of the software product" as one possible buying criteria. 51.1% would do that if there are products with the same functionality, some being "green" and others being "not green".

Comparing the results in respect to the gender, the differences between women and men are nearly the same in case of agreeing positions whereas the differences are bigger in negative responses: 11% of the men say that they cannot imagine that environmental issues of software can be a buying criterion for software while 2% of the women object to this idea.

In respect of the role of software usage, the following result occurs: Especially purchasers can imagine taking "rating of environmental issues" as a criterion while

---

[2]https://www.blauer-engel.de/en.

[3]https://www.energystar.gov/.

[4]http://tcodevelopment.com/.

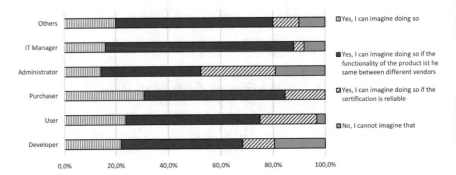

Fig. 1 Results of the questions regarding a possible selection criterion "environmental issues of software", distinguished by the role in using software, leaving out vendors and distributors due to too little respondents (Question: "Can you imagine to take 'grading of the environmental impact of the product' as a criterion while searching for a software product?")

searching for software products (Fig. 1). However, the significance of this result is quite disputable, since just 13 participants characterize themselves as being primary a software purchaser. As mentioned, the main focus group of the survey were software users. Thus, they represent the biggest group of survey participants (78.2%), followed by developers (11.5%). Between these two groups, the distribution of the answers is quite similar. However, there is a marginal tendency towards "I cannot imagine doing so" in case of developers.

According to the results, there seems to be an interest in environmental topics on the one hand (e.g. knowing eco-labels and taking them into account while buying products). However, on the other hand, these interests and awareness do not include environmental issues of software. In order to counteract here, the idea is to develop an eco-label for software. In the context of a project called "Sustainable Software Design",[5] we are currently working on such a certification.

The second part of the survey addressed the question if the ideas for criteria developed in the project arouse social interests. The survey participants were asked if the criteria "should be labeled", "should rather be labeled", "should rather not be labeled" or "should not be labeled". Overall, especially the aspects *energy efficiency* (77.5% said that it should be labeled) and *hardware efficiency* (62.5%) gain big interests. The idea to label *platform independence* and *backward compatibility* also seems to be attractive, whereas especially the criteria *hardware sufficiency* (35.4% "should be labeled") and *quality of product information* (31.6%) are deemed less important to be labeled by the survey participants. Overall, the answers to the question "Which ones of these environmental issues of software should be labeled?" show that there seems to be a general interest for the presented aspects.

---

[5]http://green-software-engineering.de/en/project/ufoplan-ssd-2015.html.

3.3 Discussion and Comparison

Comparing our survey to the other studies introduced in Sect. 2, we noticed that most of them relate eco-labels to Green IT contexts. The different studies show that many people generally know environmental labels. In our study, only 4.6% of the respondents did not know any of the three labels (ENERGY STAR, Blue Angel, TCO Certification). Indeed, knowing the certifications is just one part of the intended awareness. More important seems to be that the products and information of the labels are also known and thus recognized in the purchasing processes. We observed a lack of knowledge in the products and even more in the criteria behind the eco-labels. Ramachandiran (2012) found a similar knowledge gap about ENERGY STAR products.

In case of the Malaysian students, about half of the interviewed persons consider the ENERGY STAR while purchasing new electronic devices and computer equipment (50% of the non-ICT students agreed, 57% of the ICT students agreed Selyamani and Ahmad 2015). Ferreira (2011) states that "58% of respondents would use energy labeling in their software application selection process" (21% would not).

Indeed, so far, energy consumption of software or similar environmental aspects have not been connected to software products by the users. This is shown in the Green Software Awareness Survey by Ferreira (2011). Our results confirm this as well.

Overall, respondents state that the survey made them aware of the addressed topics (Ferreira 2011: "Until today, I have never thought much about it, but it's a very interesting subject"; similar observations in our study) and lead to further consideration of green computing issues (Pang et al. 2016). This can be seen as a first step of bringing (mainly) scientific topics to society. Further approaches into this direction will be discussed in the following section. Generally, after informing about the topics, it seems to be very important to move from knowledge to acting.

4 Approaches on How to Create Awareness

Although there seems to be a general awareness of energy consumption of software and the potential for energy savings by optimizing these products (Ferreira (2011): "98% of respondents believe that two pieces of software can have different energy consumption profiles for the same functionality", "35% of respondents consider that there's a large potential for savings. 34% of respondents consider that there's a moderate potential for savings."), neither do the programmers address nor do the user request energy efficiency to a great extent (Pang et al. 2016). Above that, "[energy] concerns are largely ignored during maintenance" (Manotas et al. 2016).

Ferreira (2011) comes up with the following hypothesis: "Given the clear awareness, respondents are taking energy consumption into consideration when buying or developing software applications." Thus, a lot of ideas on how to create this awareness have been published in the last years. They will be briefly presented in the following.

4.1 Education

According to Pang et al. (2016) the integration of green computing aspects in the education and training of programmers is just slightly implemented. Thus, one option to create a bigger awareness for green issues in the context of ICT, is to extend these topics in educational programs. Especially traditional IT projects should be addressed by these activities to show the link between algorithm and energy consumption (Pang et al. 2016) since, so far, "energy usage requirements are more common for mobile and data center projects" (Manotas et al. 2016). Next to integrating these issues into teaching and education, universities and schools could set a good example. This could support the education activities even if first studies could not show the positive effects (Selyamani and Ahmad 2015). Furthermore, the information channels primarily used by students could also be used to spread environmental IT information (Dookhitram et al. 2012). After leaving universities, the Green IT training should be provided by ICT organizations. According to Kogelman (2011) this is done in under half of the cases as stated by the surveyed ICT managers.

4.2 Labels and Information for End Users

Overall, labels like the ENERGY STAR, are mentioned quite often regarding the question how to create awareness for environmental topics (Selyamani and Ahmad 2015; Lago and Jansen 2010; Zhang et al. 2014). According to Ferreira (2011) "46% of respondents classified the impact of energy labeling as strong or very strong". Different studies point out that informing initiatives can increase the user awareness towards energy efficiency of software and consumer's positive attitude towards green products (Zhang et al. 2014; D'Souza et al. 2006). As far as we know, there are some initiatives towards labeling green software (Kern et al. 2015). However, a standardized eco-label is still missing in this context.

4.3 Recommendations and Information During Development

Since the programming phase seems to be important to monitor energy consumption (Ferreira (2011): "58% of suppliers monitor energy consumption in the programming stage"), the programmers should be informed about green computing and software issues. Pang et al. (2016) point out that "[only] 18% of the respondents take energy consumption into account during software development." According to the same study, the energy consumption is more interesting for the programmers in case of mobile development (65 versus 12% in case of software respond that it is a decision factor). So far, corresponding recommendations are missing. However, the practitioners are sure that they can learn about the improvement of the energy efficiency (Manotas et al. 2016; Pang et al. 2016). Similar to the priorities of the respondents in our survey, the functionality of the software seems to rank first for programmers (Pang et al. 2016).

4.4 Other Approaches

Lago et al. propose to do continuous measurement and feedback to bring green knowledge to researchers and practitioners. Such services could "give feedback on the carbon footprint of SBAs [service-based applications] and of their end users". A so-called "service greenery" can offer green software services via Web. However, these ideas are, so far, theoretical concepts. They need to be proven and brought to practice (Lago and Jansen 2010).

Another possibility to increase awareness for Green IT topics could be related policies (Dookhitram et al. 2012) or official legislation (Kogelman 2011). According to Kogelman (2011) the situation that European organizations did not implement Green IT practices is caused by the absence of official legislation and "no pressure from management or customers".

5 Conclusion and Outlook

Summarizing, the different studies show that the awareness for environmental issues of ICT could be higher. This is true for different roles in using hard- and software. There are approaches on how to increase the awareness. Some of them have been shortly described in this paper. Indeed, many of the proposed activities still need to be transferred from science to society. The Green IT knowledge of users, developers (including students and practitioners), and managers needs to be enlarged. They need (1) to know that ICT, including hard- and software, has environmental impacts, (2) to get information, tools, methods, and advice on how to transfer their knowledge into actions, and (3) to overall increase the awareness to these issues in their surroundings and organizations. Thus, the next steps in increasing the social interest in green computing should include activities that bring the research activities to industry, policy, education, and consumers. To do so, we will follow up on the development of an eco-label for software products, based on the results of our survey.

References

Andrae ASG, Edler T (2015) On global electricity usage of communication technology: trends to 2030. Challenges 6(1):117–157

D'Souza C, Taghian M, Lamb P (2006) An empirical study on the influence of environmental labels on consumers. Corp Commun: Int J 11(2):162–173

Dookhitram K, Narsoo J, Sunhaloo MS, Sukhoo A, Soobron M (2012) Green computing: an awareness survey among university of technology, mauritius students. In: Conference proceeding of international conference on higher education and economic development, Mauritius

Ferreira M (2011) Green software awareness survey: results. Presented at report workshop green software architecture, Tuesday 7 June 2011

GeSI, Global e-Sustainability Initiative (2015) #SMARTer2030: ICT solutions for 21st century challenges. http://smarter2030.gesi.org/downloads/Full_report.pdf

Kern E (2016) The development of an eco-label for software products—a transdisciplinary process? In: Mayr HC, Pinzger M (eds) INFORMATIK 2016: lecture notes in informatics (LNI), Bonn, pp 1285–1296

Kern E, Dick M, Naumann S, Filler A (2015) Labelling Sustainable Software Products and Websites: Ideas, Approaches, and Challenges. In: Johannsen VK, Jensen S, Wohlgemuth V, Preist C, Eriksson E (eds) Proceedings of EnviroInfo and ICT for Sustainability 2015: 29th international conference on informatics for environmental protection (EnviroInfo 2015) and the 3rd international conference on ICT for sustainability (ICT4S 2015). Copenhagen, September 7–9, 2015. Atlantis Press, Amsterdam, pp 82–91

Kogelman C-A (2011) CEPIS Green ICT survey—examining green ICT awareness in organisations: initial findings. Carol-Ann Kogelman on behalf of the CEPIS Green ICT Task Force. CEPIS UPGRADE XII(4):6–10

Lago P, Jansen T (2010) Creating environmental awareness in service oriented software engineering. In: International conference on service-oriented computing, pp 181–186

Malmodin J, Bergmark P, Lundén D (2013) The future carbon footprint of the ICT and E&M sectors. In: Hilty LM, Aebischer B, Andersson G, Lohmann W (eds) ICT4S ICT for sustainability: proceedings of the first international conference on information and communication technologies for sustainability, February 14–16, 2013. ETH Zurich, University of Zurich and Empa, Swiss Federal Laboratories for Materials Science and Technology, Zürich, pp 12–20

Manotas I, Bird C, Zhang R, Shepherd D, Jaspan C, Sadowski C, Pollock L, Clause J (2016) An empirical study of practitioners' perspectives on green software engineering. In: Proceedings of the 38th international conference on software engineering, pp 237–248

Pang C, Hindle A, Adams B, Hassan AE (2016) What do programmers know about software energy consumption? IEEE Softw 33(3):83–89

Ramachandiran CR (2012) Green ICT practices among tertiary students: a case study. In: Business, engineering and industrial applications (ISBEIA). IEEE, pp 196–201

Selyamani S, Ahmad N (2015) Green computing: the overview of awareness, practices and responsibility among students in higher education institutes. J Inf Syst Res Innov

Zhang C, Hindle A, German DM (2014) The impact of user choice on energy consumption. IEEE Softw 31(3):69–75

A Framework for Optimizing Energy Efficiency in Data Centers

Volkan Gizli and Jorge Marx Gómez

Abstract Recent studies have shown that around 50,000 data centers are causing 2% (10 TWh) of the whole power consumption in Germany. Despite the numerous efforts made to improve energy efficiency, according to the forecast for 2020 expected total power consumption is going to be around 14 TWh. Because of this, a grey energy (indirect energy, which is used for production, transport, storage, sale, and disposal) needs to be considered for an overall energy management optimization. The Carl von Ossietzky University of Oldenburg is addressing this problem in the research project called TEMPRO (Total Energy Management for Professional Data Centers). Moreover, as part of the project, a framework is proposed, which will allow a preliminary assessment of data centers' energy efficiency through a depicted visualization that considers a conformity testing of certificated KPIs. Further, the framework will emit guidance that are going to be applied to optimize energy efficiency.

Keywords KPI · Energy efficiency · PUE · Data centers

1 Introduction

Recent studies have shown that around 50,000 data centers are causing 2% (10 TWh) of the whole power consumption in Germany (TU Berlin IZE 2008). Despite the numerous efforts made to improve energy efficiency of data centers, expected forecast of a power consumption for the year 2020 is around 14 TWh (Quandt 2017). And one of the main drivers is increase of a main memory (e.g. DRAM) adoption, which itself depends on a storage and processing capacities. For instance, in the 2005 power consumption of a storage systems was 5% of the total data

V. Gizli (✉) · J. Marx Gómez
Carl von Ossietzky Universität Oldenburg/VLBA, Oldenburg, Germany
e-mail: volkan.gizli@uni-oldenburg.de

J. Marx Gómez
e-mail: jorge.marx.gomez@uni-oldenburg.de

© Springer International Publishing AG 2018
B. Otjacques et al. (eds.), *From Science to Society*, Progress in IS,
DOI 10.1007/978-3-319-65687-8_24

centers power consumption (Koomey 2008), whether in the 2011 it raised to the 17% (Prakash et al. 2014). In general, when optimizing IT components in data centers, the power usage effectiveness (PUE) value usage as a reliable metric degrades, because the PUE value does not provide enough information about the data center and also able to present part of reality. Thus, a grey energy, which can be defined as an energy used for aspects such as production, transport, storage, sale, and disposal, should be considered for more realistic energy management optimization (Prakash et al. 2012). As a consequence of a rapidly increasing growth of constituents in data centers such as servers, storage facilities and network components, the topic of energy management optimization is becoming important. However, many data centers haven't perform any actions in order to develop those optimization techniques and solutions that supposed to increasing energy efficiency within a particular data center. The primary reasons are criteria that are not transparently determine energy potentials. For instance, one of such reasons can be the bad quality of a various Key Performance Indicators (KPIs), which are used for such type of measurements.

As it was mentioned before, the primary KPI for measuring energy efficiency is the power usage effectiveness (PUE). Popularity of this measurement for understanding energy consumption in data centers can explained by its simplicity (e.g. lower is better). Many companies use the PUE as a generic KPI to demonstrate how "green" their data centers are (Klingert et al. 2013). The PUE indicates total energy consumption in ratio to the internal consumption of data centers. It means that as long as the value converge to 1, the better situation with energy consumption is Reisinger (2014). For example, in case a value of the PUE is 3 it will mean that two-thirds of the power is used by the heat dissipation and just one-thirds of overall power is real consumed by the technical equipment, which is not a good sign. On the other hand, the PUE value of 1.3 is distinguished. Such value means that 30% of the power is used inefficient. The PUE value of 1 is ideal and means that no more power is necessary for the hardware (ITWissen 2016). The PUE value under 1 will be profitable. Such value can be reached, in case it is possible to use generated heat to spare heating costs of a data center. Conversely, this means that a total of the PUE value will be below 1.

The Carl von Ossietzky University of Oldenburg is trying to overcome this problem in a research project TEMPRO (Total Energy Management for Professional Data Centers), which is done in a cooperation with number of partners who are trying to optimize their own data centers by directly applying methods and techniques developer within the project.

2 Research Project TEMPRO

The research project TEMPRO (Total Energy Management for Professional Data Centers; funded by Federal Ministry for Economic Affairs and Energy) pursues an improvement of total energy management for professional data centers. The Carl von Ossietzky University of Oldenburg as a partner of this research project is located in the northwestern part of Germany. The university research group is represented by three following subgroups:

- Cascade Use Gruppe (CCU)
- Very Large Business Applications (VLBA)
- Innovationsmanagement und Nachhaltigkeit (PIN)

Further, academic partners of the research project TEMPRO are:

- Technical University of Hamburg-Harburg
- Borderstep GmbH

Furthermore, company partners of TEMPRO are:

- BTC IT Services GmbH
- KDO (Zweckverband Kommunale Datenverarbeitung Oldenburg)
- b.r.m (business resource management) Technologie- und Managementberatung
- dc-ce Berlin-Brandenburg GmbH
- Mairec Edelmetallgesellschaft GmbH
- CEWE Stiftung & Co. KGaA
- Hewlett Packard Enterprise Company
- e3 computing GmbH.

2.1 TEMPRO Work Packages

Due to the direct connection of the TEMPRO research project to the problems of energy consumption optimization in data centers and in which area the following approach is part of will be presented more detailed. The research project TEMPRO has totally five work packages. The following picture shows all work packages, which includes the work package 3, which is dedicated for the subgroup Very Large Business Applications (VLBA) of the University of Oldenburg.

The title of the work package 3 is "Information and Evaluation Methods for Energy Efficiency in Data Centers", which includes three sub work packages (Fig. 1).

2.2 TEMPRO Work Package 3

The Fig. 2 shows three sub work packages of the TEMPRO's work package 3.

The work package 3.1 addresses an issue of design and development of a new sort of KPIs, which are based on the existing KPIs. However, newly developer KPIs are going to take into consideration grey energy and other resources. In order to

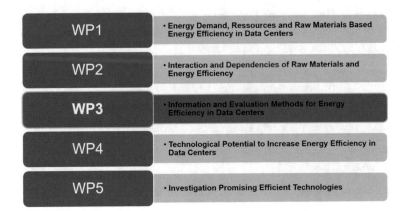

Fig. 1 TEMPRO work packages

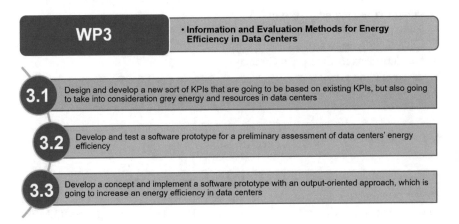

Fig. 2 TEMPRO sub work packages 3

accomplish mentioned goal, existing KPIs, which are used as an indicator for energy efficiency in data centers for a moment, are going to be analyzed and justified. Based the analysis results, additional KPIs and indicators going to be developed. These additional KPIs should be assessed and tested especially with focus on aspects of a grey energy and resources. This is going to build a basics for the further development and implementation of a software prototype that will reflect an assessment model and is going to be implemented in the work package 3.2 and 3.3.

The work package 3.2 is dedicated to develop a software prototype for a preliminary assessment of data centers' energy efficiency. For the assessment of the special aspects such as data center types, functions and features, IT equipment, cooling type and many others are going to be used.

The work package 3.3 is dedicated to develop a concept and implement a software prototype based on the results obtained in work package 3.1 and 3.2,

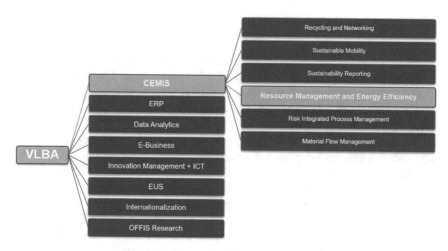

Fig. 3 VLBA research fields

which is going to increase the energy efficiency in data centers. As it was mentioned before, this work package accumulates results produced from the work packages 3.1 and 3.2.

2.3 TEMPRO in the Research Field of VLBA

In the context of the VLBA research fields the work package 3 "*Information and Evaluation Methods for Energy Efficiency in Data Centers*" is adapted under the field "*Resource Management and Energy Efficiency*", which is embedded again under the field CEMIS (Corporate Environment Information Management Systems). CEMIS is an organizational-technical system for systematically obtaining, processing and making environmentally relevant information available in companies (Rautenstrauch 2013).

The Fig. 3 shows the structure of the VLBA research fields.

3 TEMPRO Analytics

As an approach for an optimizing of an energy efficiency in data centers a framework called TEMPRO Analytics is proposed. This framework is going to allow a preliminary assessment of data centers' energy efficiency and inform the supervisor about elements/components of an infrastructure that may cause possible problem and become a bottleneck in energy consumption.

3.1 Framework System

As the first step in the development of the framework TEMPRO Analytics it is necessary to identify suitable indicators for the assessment. The framework will allow to apply a set of chosen indicators to a data center's infrastructure, in order to allow monitoring of indicators and infrastructure as well be able to benchmark the current status.

The certification "Der Blaue Engel" (Egovernment-Computing.de), a German eco-label, has the goal to optimize the energy efficiency in data centers. Taking into consideration experience provided by KPIs that used as a part of "Der Blaue Engel" certification, it is important to consider this certification within the benchmark. The overall objective of the platform is to propose an action in case current status does not comply with the standard measured values. These actions will inform the supervisor about elements of the infrastructure that may cause some problem.

The framework will provide different means of visualization, like Sankey diagrams (Sankey-Diagrams.com) for the preliminary assessment. In addition, it will allow to connect different data sources. Besides that, it will offer an editor in order to simplify customization of performance indicators. TEMPRO Analytics will make use of XBRL (Bergon 2004) as reporting language.

This framework has to meet following requirements:

1. Using chosen KPIs for data centers

   - KPIs of "Der Blaue Engel" as a certified benchmark
   - Further relevant KPIs for data centers

2. Visualization of data centers' specific KPIs

   - Preliminary assessment of energy efficiency
   - Visualization (e.g. Sankey diagrams)

3. Certification or conformity testing based on chosen KPIs
4. Guidance based on conformity testing outcomes
5. Allows to connect different data sources
6. Offers an customizing editor for performance indicators
7. Make use of XBRL as reporting language (Bergon 2004).

3.2 Example

Here is an example, if the KPI for the coaling in a data center has a certain value, which is higher than the defined value of the certificate "Der Blaue Engel":

1. Visualizing certain values in a data center

   - For example coaling values
   - Depict values with Sankey diagrams

2. Coaling value higher by value X

   - Certification/conformity testing (comparison calculation)
   - Done by using defined KPIs of "Der Blaue Engel"

3. Guidance

   - Decrease the value by value Y in the affected area, whereby no difference will occur
   - Saving of energy.

4 Conclusion

In general, the framework approach TEMPRO Analytics will allow a preliminary assessment of data centers' energy efficiency through a depicted visualization with consideration on a conformity testing of certificated KPIs. The resulting guidance are going to be applied in a certain areas of data centers to optimize overall energy efficiency. In addition, proposed approach will provide different means of visualization, like Sankey diagrams, for a preliminary assessment. Moreover, the framework can be further enhanced with advanced features, which may not existing at the moment. For example, it is possible to combine different data sources together and also offer a performance indicators editor-tool in order to simplify customization process. Beside these aspects, the proposed framework should be able to take into consideration further important KPIs in order to continuously optimize the energy efficiency.

Acknowledgements This work is part of the project TEMPRO (Total Energy Management for Professional Data Centers) with a financial support by the Federal Ministry for Economic Affairs and Energy of Germany (grant number: 03ET1418A-H).

References

Abolhassan F (2013) Der Weg zur modernen IT-Fabrik: Industrialisierung—Automatisierung—Optimierung. Springer Fachmedien Wiesbaden

Bergeron B (2004) Essentials of XBRL: financial reporting in the 21st century. Wiley, New York

Egovernment-Computing.de (Online). http://www.egovernment-computing.de/blauer-engel-fuer-betreiber-von-rechenzentren-a-444734/. Accessed 15 Jan 2017

ITWissen (2016) PUE (power usage effectiveness), PUE-Wert. http://www.itwissen.info/definition/lexikon/power-usage-effectivness-PUE.html. Accessed 20 Feb 2017

Klingert S, Hesselbach-Serra X, Ortega MP, Giuliani G (2013) Energy-efficient data centers: second international workshop E2DC 2013, Berkeley, CA, USA, 21 May 2013. Revised selected papers. Springer, Berlin

Koomey JG (2008) Worldwide electricity used in data centers. Environ Res Lett 3(3):034008

Prakash S, Baron Y, Ran L, Proske M, Schlösser A (2014) Study on the practical application of the new framework methodology for measuring the environmental impact of ICT—cost/benefit analysis. European Commission, Brussels, Studie

Prakash S, Liu R, Schischke K, Stobbe PL (2012) Early replacement of notebooks considering environmental impacts. In: Proceedings electronics goes green 2012. Berlin, pp 1–8

Quandt R (2017) Google baut neues Rechenzentrum in Holland für 600 Mio. Euro (Online). http://winfuture.de/news,83744.html. Accessed 15 Jan 2017

Rautenstrauch C (2013) Betriebliche Umweltinformationssysteme: Grundlagen, Konzept und Systeme. Springer-Lehrbuch, Springer, Berlin

Reisinger N (2014) Green-IT-Strategien für den Mittelstand: Nachhaltige Lösungen in der IT und durch IT-Unterstützung. Nachhaltigkeit Series. Diplomica Verlag

Sankey-Diagrams.com (Online). http://www.sankey-diagrams.com/sankey-diagram-software/ Accessed 15 Jan 2017

TU Berlin IZE (2008) Konzeptstudie zur Energie- und Ressourceneffizienz im Betrieb von Rechenzentren. Studie zur Erfassung und Bewertung von innovativen Konzepten im Bereich der Anlagen-, Gebäude und Systemtechnik bei Rechenzentren. TU Berlin—Innovationszentrum Energie (IZE), Berlin, 2008. in der IT und durch IT-Unterstützung. Nachhaltigkeit Series, Diplomica Verlag, 2014

GranMicro: A Black-Box Based Approach for Optimizing Microservices Based Applications

Ola Mustafa, Jorge Marx Gómez, Mohamad Hamed and Hergen Pargmann

Abstract Microservices application architecture emerged as a special application architecture of Service Oriented Architecture (SOA). It is also known as cloud native applications architecture. For component based web applications, a transformation to microservices architecture requires an extensive refactoring process to be evolved to microservices. Current practice for the main activity in this refactoring process, is following the white box based approaches which require enormous time to analyze all business processes in the business domain besides not considering nonfunctional requirements such as performance and resource consumption. This paper explains the design and the validation of GranMicro as a black box based approach, applied on a sample browsing scenario in e-bookshop web application. The aim of this approach is to support the decision of a main design issue in this software architecture which is the service granularity. This decision usually relies on the experience of the software architect rather than a structured approach, which is the main contribution of this paper.

O. Mustafa (✉) · J.M. Gómez · M. Hamed
Department of Computing Science, Business Informatics,
University of Oldenburg (UOL), Oldenburg, Germany
e-mail: ola.mustafa@uni-oldenburg.de

J.M. Gómez
e-mail: jorge.marx.gomez@uni-oldenburg.de

M. Hamed
e-mail: mohamad.hamed@uni-oldenburg.de

H. Pargmann
Jade Hochschule, Wilhelmshaven, Germany
e-mail: hergen.pargmann@jade-hs.de

© Springer International Publishing AG 2018
B. Otjacques et al. (eds.), *From Science to Society*, Progress in IS,
DOI 10.1007/978-3-319-65687-8_25

283

1   Introduction

Service based applications and microservices are widely adopted by e-commerce applications, motivated by the high flexibility for this special type of application architecture, and its ability to adapt in front of extremely high workloads, that's why this architecture considered to be a native architecture for cloud based applications (Mustafa and Marx Gómez 2016a). In the last few years, very large e-commerce applications announced the migration of their web applications architecture to microservices, to gain the benefits of higher performance and scalability besides better resource utilization over the cloud (Villamizar et al. 2015).

The main idea of this architecture is not only to isolate the business layer from the presentation layer, but also to split the business layer into a group of small services (microservices), those of which are triggered from one single Interface called the application gate way. The transformation of an existing web application to microservices architecture implies the process of splitting the application code to a group of small web services, so that each service contains a unique part of the business processes, this part should be highly decoupled and isolated from other parts (Villamizar et al. 2015).

Figure 1 shows two different web application architectures; the left one follows the component based architecture, where the whole code is released as one large body no matter how it is arranged inside (factions, classes, libraries, etc.). On the other hand, microservices architecture is shown on the right side where the application is divided into small isolated services all are communicated through the main interface (application gateway).

1.1   Service Granularity as a Design Issue

The question of how big the service should be or in other words, how many business processes each web service should encapsulate, is a main design issue which is known as service granularity (Kulkarni and Dwivedi 2008). That is because low granularity (which implies very small and high number of services) will increase the complexity of managing those services. However, high granularity (which implies bigger services and few number of services) will not achieve the desired economic design over the cloud, where resource consumption is calculated per use.

Moreover, the current practice to reach the optimal level of service granularity is following a multi iteration process. This process driven by the results of monitoring the deployed services, results in terms of performance and resource consumption. Figure 2 shows the iterative process until reaching the optimal level of service granularity in microservices architecture.

The main problem discussed in this paper, is how to come over this iterative process by trying to plan for the optimal granularity level early in the development

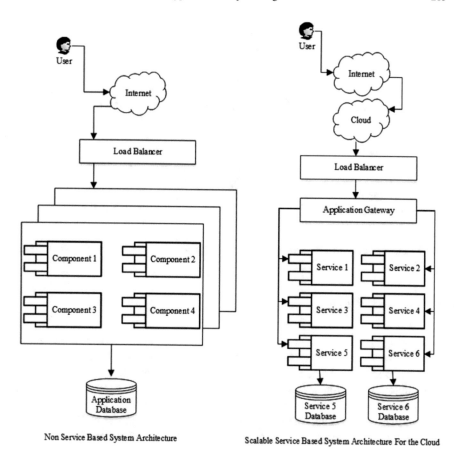

Fig. 1 Component based web application architecture (*left*) versus Microservices based architecture (*right*) (Mustafa and Marx Gómez 2016b)

life cycle. The action of this early planning will save a lot of time and will avoid nonfunctional requirements leaks as will be explained in the validation section.

This paper lies in four main parts; the first is to discuss some related work, those of which mentioned microservices architecture design and the way taken toward it. This part aims to spot the light on the gap filled by this research. The second part discusses the general proposed method, the motivation behind it and the main objective from it. The third part discussing the core concept of GranMicro in details, applied on the browsing scenario in a sample e-bookshop web application. The final part of this paper is discussing the validation of the proposed method, this is explained by comparing the impact of deciding the granularity when GranMicro is applied and when it is not applied.

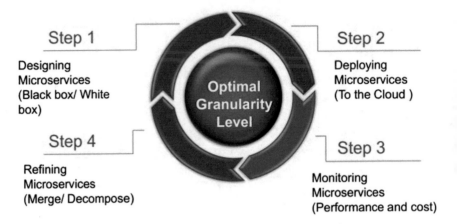

Fig. 2 The iterative process towards the optimal granularity level for microservices based applications

2 Related Work

Refactoring applications to microservices architecture, would require a decomposition phase, to break the architecture into loosely coupled services. The objective from discussing the related work is to analyze the strategies taken to decompose the monolithic architecture. The discussed papers are those from the literature, re-factored the monolithic architecture to microservice architecture, driven by the need to evolve into cloud native applications. This is due to the main features of microservice architecture which are mainly the effective scalability in front high workload and its economic consumption for hardware resources (Villamizar et al. 2015).

The experience report (Chauhan and Babar 2011), mentioned the requirements for scalability, which considered to be fulfilled by applying the stateless design of services and REST. The decomposition phase is not following a structured approach. The empirical evaluation experiment (Villamizar et al. 2015) evaluated the microservice architecture under high loads, compared the generated cost with the monolithic architecture under the same load. The better performance and the cost efficient of resource utilization concluded for the microservice architecture. Both architectures tested under the vertical scaling method only, and the decomposition phase is also not following a structured approach. In Balalaie et al. (2015), the author mentioned the microservices architecture as a novel architecture and considered it as the cloud native architecture. The scalability needs on the application level addressed, fulfilled by refactoring the application into service based architecture, and the decomposition is done based on domain entities. The report did not mention any tests to evaluate application scalability after migration. Development and deployment complexities were the main challenges expressed. In Gupta (2014), the author proposed a recursive approach for decomposing the business logic into web services and the metric of loose coupling was the only one mentioned. The need for an adaptive architecture to scale addressed, to serve the increasing demand on peak loads. In Kwon and Tilevich (2014), a method for model based automated transitioning the monolithic architecture into objects and services was proposed, the refactoring motivated by the need to improve the performance efficiency, by processing large volumes of data in parallel. Scaling infrastructure resources was not mentioned. For identifying services, a recommendation engine built based on applying clustering techniques with loose coupling metrics, services decomposed on class level.

The previous mentioned papers and reports, are divided into two main groups, the first group evolved to microservice architecture without following a clear structured approach for the decomposition phase, the second group followed the white box based approach and started from a business model. Such approach would not consider any nonfunctional requirements in the design phase and focusing on decoupling the services as much as possible.

The researchers in Gysel et al. (2016), referred to the issue where most references about microservices explained the decomposition phase as an art and totally depends on the experience of the software architect. Based on that, the authors proposed a tool for decomposing microservices (Service Cutter) with a basic concept of applying different cluster based algorithms on the business process model. The tool is said to be the first structured approach for decomposing microservices.

The approach proposed in this paper aims to bridge this gap by providing a supportive approach for software architects helping in making optimized granularity decisions for microservices architecture besides considering the nonfunctional requirements.

3 Proposed Approach: GranMicro

In this section, GranMicro approach is proposed and explained as a black box based approach targeting software architects, helping them to decide the level of granularity for any new microservices architecture evolved from an existing component based web application. In a previous paper research (Mustafa and Marx Gómez 2017), the authors provided an experimental study and argued for the impact of having different service granularities for the same microservices based web application running in a cloud environment. The main impact was discussed on the response time and on the CPU consumption. The results recorded different behaviors for different granularity levels over the same investigated scenario. Based on that, this experiment motivated the need of GranMicro as a black box based approach, aims to help defining the most feasible granularity level for microservices based application.

The main concept of GranMicro approach is based on extending the utilization of web usage mining techniques and leverage it to the level of optimizing back end design. This is based on extracting workload pattern from web access logs and build the decision of decomposition based on the generated pattern. The main input for GranMicro is the web access logs and the main output is the service model diagram.

3.1 Workload Characterization

Investigating the impact of workloads over microservices based applications stared to activate lately (Ueda et al. 2016). However, extending the web usage mining techniques to optimize the granularity of microservices is not mentioned yet. Domain Driven Design is the main base for all granularity based decisions, where the microservices are splitted based on the application domain. This approach will be enough to generate the basic level of granularity but not the optimized one. This is due to that business domain analysis may not reveal some high loaded parts of the running system.

Mining web access logs to extract useful information about user behavior has been extensively investigated in research (Ansari et al. 2015). The mining process goes through main steps to clean the logs and prepare them for sessionizing, this is the main process where the users are distinguished based on the IP address and the time stamp over different web pages of study. Different researches applied different clustering approaches on the huge logs, in order to extract the usage pattern and then making adaptive decisions. Such decisions were focusing on enhancing the user experience such as, designing recommender systems or adapting the design of the interface.

In the same domain of web usage mining, few work extended usage pattern extraction to the level of workload pattern extraction. Even though, such extensions were not focused on adapting the application architecture, but rather to generate a workload models as input for simulated load testing as in Abbors et al. (2014).

In the common practice of web usage mining, the main distribution of user sessions and web pages is the base to generate the user's navigation pattern. This approach is very useful when the target is to enhance the user experience. However, it could be reformed to extract the workload distribution pattern between the web pages, to see which pages are attracting the high loads within certain periods of time (time slots). This reformation requires that the main sessionizing phase to be based on time stamps rather than on user sessions (IP addresses), as in Fig. 3.

The next step is to decide how to cluster time periods into (peaks), to decide which pages attract high loads or user visits. Here where the application of fuzzy based clustering sounds proper and serving the need as will be clarified in the next section.

	Time	IP	URL	Ref				
	0:01	1.2.3.4	A	_	Session 1	0:01	1.2.3.4	A
	0:09	5.6.7.8	B	A		0:09	5.6.7.8	B
Time Slot 1	0:19	4.3.2.1	C	A		0:19	4.3.2.1	C
	0:25	8.7.6.5	E	C		0:25	8.7.6.5	E
	1:26	1.3.2.4	A	_	Session 2	1:26	1.3.2.4	A
	1:28	5.7.6.8	F	C		1:28	5.7.6.8	F
	1:30	4.2.3.1	B	A		1:30	4.2.3.1	B
	1:36	8.6.7.5	D	B		1:36	8.6.7.5	D

Fig. 3 Time slot based sessionization (Mustafa and Marx Gómez 2016a)

3.2 Extracting Workload Pattern as Peaks

Among many usage mining approaches (Shi 2009; Ansari et al. 2015; Anari et al. 2015), unsupervised methods, especially clustering, used to group users with common browsing behavior. However, while k-means clustering used for huge access logs because of its computational efficiency (Elhiber and Abraham 2013), it still suffers from the absence of possibility to obtain overlapping clusters with current approaches, so that a user can belong to more than one group (Boob and Dakhane). Fuzzy clustering on the other hand, could deal with the ambiguity and the uncertainty underlying web access data, this is by assigning weights to URLs and user sessions based on a fuzzy membership function.

With session weight assignment approach, time slot based sessions are proposed to weight the sessions over the time stamps rather than the users. After assigning the weights we move to the step of applying a "Fuzzy c-Mean Clustering" algorithm to discover the clusters of peak periods rather than user profiles. Figure 4 shows how the weight assignment should be applied over the time stamps extracted from the logs, in order to extract the main peaks.

This approach allows generating peaks oriented sessions rather than user oriented ones. Which also allows the observation of workload distribution over the

	Cat A	Cat B	Cat C	Cat D
Peak 1	1	1	2	4
Peak 2	2	1	3	4
Peak 3	1	0	4	4
Peak 4	2	0	4	4
Peak 5	1	0	3	3
Peak 6	1	1	3	3
Peak 7	1	1	3	3
Peak 8	1	0	2	2

Fig. 4 Time slot based sessionization leads to clustering over web pages categories (Mustafa and Marx Gómez 2016b)

web pages. Finally, those web pages with (higher) workloads, should have the priority to be splitted to microservices. Figure 5 explains the steps of GranMicro approach where initially starts with extracting the web access logs from the web server, then preparing them by removing the noise and completing the broken paths, after that sessionizing the records according to the time stamps and then clustering them to generate peaks. Then, web pages categories will be classified according to the high loads and the highest loaded pages could then assigned for the highest priority of splitting as in the last step.

4   GranMicro Implemented on Micro-BookShop

Micro-BookShop is an online book shop, providing the main e-shop functional requirements (browse, search, order) as well as the main shop-cart features. It is a part of an ongoing research, aims to investigate the impact of service granularity on the performance of microservices applications over the cloud. Moreover, the impact of considering the workload pattern and deciding the level of granularity on having economical performance on the cloud, based on the results.

The presentation layer for Micro-BookShop web application is implemented using Angularjs, while the backend of service layer is implemented with VS.net 2015 (web api/c#). Micro-BookShop is deployed on cloud azure, the database used for this experiment is MS SQL Server 2016. The data are real, and scraped from amazon.com with 5000 records. Finally, all load testing for this experiment implemented using visual studio team services on azure cloud and the workload generated automatically using the technique of record and play.

Micro-BookShop considered to be a good example to apply GranMicro for two reasons. The first, is because this microservices architecture evolved from a component based application in the first place. The second, is the motivation for this evolvement was the high cpu consumption recorded for the browsing scenario for the load test with 15000 user/min.

The main aim of applying GranMicro on this scenario is to observe the impact of it on the CPU consumption and the response time as a main indicator of performance. However, microservices is already known as scalable architecture for high workloads (Villamizar et al. 2015). In other words, adopting microservices generally is expected to improve the mentioned benchmarks anyway. Based on that, Microservices design will be applied twice; one time without GranMicro and the other design will be with appling GranMicro.

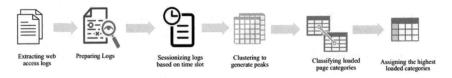

Extracting web   Preparing Logs   Sessionizing logs   Clustering to   Classifying loaded   Assigning the highest
access logs   based on time slot   generate peaks   page categories   loaded categories

Fig. 5 GranMicro approach

4.1 Browsing Scenario Under Load Testing

The browsing scenario in Micro-Bookshop starts from the home page, where the user starts navigating for books from the menu then, by selecting certain book the user will be transferred to another page which retrieves all book details for him. The details page is divided into three parts; the first part where main details about the book is provided like image, title, price, etc. and a main link to transfer the user to his cart, in case he would like to purchase the book. The second part is the reviews, where all reviews about the book are gathered from previous customers, here it is important to mention that not all reviews are retrieved but only the last ones, in the bottom of this section a link to the full reviews is provided. The third part in the page is the recommendations part, where a group of recommended books are retrieved as suggestions for the user to browse. Following the GranMicro approach, if the web logs extracted, prepared and processed, then followed by a time slot based sessionization and then the clustering and the classification applied, the results shown to be as in Fig. 6. The table shows the workload distribution over three different page categories, all of which could be reached from the book details page. The results shows in average that the highest load recorded to be for the reviews pages followed by the recommendations URLs which are calculated by collecting the referral links to book details page from the same page not from the home page.

According to GranMicro approach, book reviews pages should have a higher priority to be splitted on two separated web services (microservices), more than the other categories. Based on that, we will have two different versions for evolving the browsing scenario to microservices as in Fig. 7; the first version is by not following GranMicro approach and splitting the recommendations service into separated microservice, this will include encapsulating all classes and methods for this functionality into isolated web service. The second design will be by following GranMicro approach and splitting the reviews pages into separated web service.

	Reviews page	Cart page	Recomm.
Peak1 (5000 U/M)	50%	20%	30%
Peak2 (10000 U/M)	38%	18%	44%
Peak3 (15000 U/M)	52%	23%	25%
AVG.	46.6%	20.3%	33%

Fig. 6 Workload percentage classified over categories

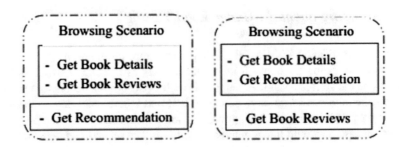

Fig. 7 Two different browsing scenarios applying microservices design; (*to the right*) where GranMicro applied

5   GranMicro Validation Results

To validate GranMicro approach, the previous two designs are load tested, each load test is implemented three times, each time with different number of users (workload). The benchmarks of CPU utilization and response time are observed for all workloads to be finally compared. The load testing will be generated three times with 5000, 10000 and 15000 user/min.

Figure 8—the left part—shows the improvement in response time for the second version where GranMicro applied, it is also noticed that that the greatest improvement in response time recorded for the highest workload. The right part shows the improvement in CPU utilization for version 2 where GranMicro applied.

The improvement in the performance and the resource consumption is the situation of optimization in performance which this research aims to reach.

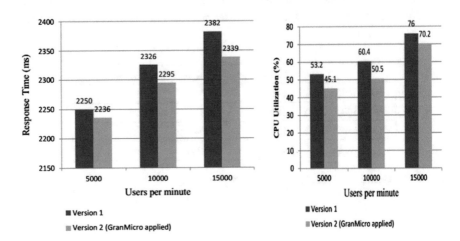

Fig. 8 Response time and CPU utilization in milliseconds for the two versions of the browsing scenario under load testing

6 Conclusion

The importance of this approach lies in two main strength points; the first one is that this approach is not starting from any business domain model (white box technique), which makes it practical and easy to apply, the only assets required is the web logs. The second point is that it will prioritize the decomposition action which will lead to decomposing only loaded services, as a result, critical non-functional requirements, as performance leaks and over resource utilization would be avoided. Based on that, GranMicro proposed in this paper as a supportive approach for software architects helping optimize microservices design.

The future work is to implement the proposed approach and convert it to hard coded then to apply it on large web application.

References

Abbors F, Truscan D, Ahmad T (2014) An automated approach for creating workload models from server log data. ICSOFT EA

Anari B, Meybodi MR, Anari Z (2015) Clustering web access patterns based on learning automata

Ansari Z, Ahmed W, Azeem M, Babu AV (2015) Discovery of web usage profiles using various clustering techniques. arXiv preprint arXiv:1509.00692

Balalaie A, Heydarnoori A, Jamshidi P (2015) Migrating to cloud-native architectures using microservices: an experience report. arXiv preprint arXiv:1507.08217

Boob MAN, Dakhane D Mining usage profiles using fuzzy clustering and its applications

Chauhan MA, Babar MA (2011) Migrating service-oriented system to cloud computing: an experience report. In: 2011 IEEE international conference on cloud computing (CLOUD). IEEE

Elhiber MHA, Abraham A (2013) Access patterns in web log data: a review. J Netw Innov Comput 1(2013):348–355

Gupta S (2014) A heuristic for architecting cloud applications. SIGSOFT Softw Eng Notes 39 (5):1–3

Gysel M, Kölbener L, Giersche W, Zimmermann O (2016) Service cutter: a systematic approach to service decomposition. In: Aiello M, Johnsen EB, Dustdar S, Georgievski I (eds) Service-oriented and cloud computing: 5th IFIP WG 2.14 European conference, ESOCC 2016, Vienna, Austria, 5–7 Sept 2016, Proceedings. Springer International Publishing, Cham, pp 185–200

Kulkarni N, Dwivedi V (2008) The role of service granularity in a successful SOA realization a case study. In: IEEE congress on services-part I, 2008. IEEE

Kwon Y-W, Tilevich E (2014) Cloud refactoring: automated transitioning to cloud-based services. Autom Softw Eng 21(3):345–372

Mustafa O, Gómez JM (2016a) SPE-SOA: towards resource consumption-aware migration to cloud based applications. In: EnviroInfo2016. Shaker Verlag, Berlin, Germany

Mustafa O, Gómez JM (2016b) Using fuzzy clustering techniques to improve the design of microservices web applications. In: Eureka OPTISAD 2016, Mexico

Mustafa O, Gómez JM (2017) Sustainable approach for improving microservices based web applications—economic perspective. In: International conference on sustainability and environmental management, Legon, Ghana

Shi P (2009) An efficient approach for clustering web access patterns from web logs. Int J Adv Sci Technol 5(1)

Ueda T, Nakaike T, Ohara M (2016) Workload characterization for microservices. In: 2016 IEEE international symposium on workload characterization (IISWC)

Villamizar M, Garces O, Castro H, Verano M, Salamanca L, Casallas R, Gil S (2015) Evaluating the monolithic and the microservice architecture pattern to deploy web applications in the cloud. In: 2015 10th computing Colombian conference (10CCC). IEEE

Part VI
Sustainable Mobility

Providing a Sustainable, Adaptive IT Infrastructure for Portable Micro-CHP Test Benches

Dominik Schöner, Richard Pump, Henrik Rüscher, Arne Koschel and Volker Ahlers

Abstract During the transition from conventional towards purely electrical, sustainable mobility, transitional technologies play a major part in the task of increasing adaption rates and decreasing range anxiety. Developing new concepts to meet this challenge requires adaptive test benches, which can easily be modified e.g. when progressing from one stage of development to the next, but also meet certain sustainability demands themselves. The system architecture presented in this paper is built around a service-oriented software layer, connecting a modular hardware layer for direct access to sensors and actuators to an extensible set of client tools. Providing flexibility, serviceability and ease of use, while maintaining a high level of reusability for its constituent components and providing features to reduce the required overall run time of the test benches, it can effectively decrease the CO_2 emissions of the test bench while increasing its sustainability and efficiency.

Keywords Battery electric vehicles · Portable Micro-CHP unit · Test bench · SOA · Adaptive IT infrastructure · Sustainable software engineering

1 Introduction

In 2007, the German government published their Integrated Energy and Climate Programme (IKEP) (Bundesministerium für Wirtschaft und Energie 2007), defining its climate protection goals until 2020, which describes electric mobility as a factor in the process of reaching those goals. The target of one million electric vehicles on

D. Schöner (✉) · R. Pump · A. Koschel · V. Ahlers
Faculty IV, Department of Computer Science, University of Applied Sciences and Arts Hannover, Ricklinger Stadtweg 120, 30459 Hannover, Germany
e-mail: dominik.schoener@hs-hannover.de

H. Rüscher
Faculty II, Department of Mechanical Engineering, University of Applied Sciences and Arts, Bismarckstraße 2, 30173 Hannover, Germany

© Springer International Publishing AG 2018
B. Otjacques et al. (eds.), *From Science to Society*, Progress in IS,
DOI 10.1007/978-3-319-65687-8_26

German roads by 2020 (Presse- und Informationsamt der Bundesregierung 2013) meanwhile was—as of 1. January 2016—still 844,133 vehicles away, according to the Kraftfahrt-Bundesamt (German Federal Motor Transport Authority) (Kraftfahrt-Bundesamt 2016), resulting in an adoption rate of 0.3% for electric and hybrid passenger cars.

To alleviate some of the issues of range anxiety in battery electric vehicles (BEV), transitional concepts combining electric motors with internal combustion engines have been developed. Hybrid electric vehicles for example use their battery for assistive purposes only and charge it via recuperation. Range extenders meanwhile are small-scale combustion engines integrated into BEV to reduce charge depletion rates, but—unlike hybrid vehicles—are usually not capable of providing enough energy for propulsion without battery support.

Since 2013 an interdisciplinary research project at the University of Applied Sciences and Arts Hannover is studying a new concept evaluating the scalability of mobile micro combined heat and power units (Schmicke et al. 2014). The goal is to replace conventional range extender with a unit scaled in accordance with the BEV's thermal requirements. This allows the cogeneration of heat and power by using the thermal output of the combustion engine as well as its torque, increasing fuel efficiency and reducing CO_2 emissions compared to other range extender concepts (Minnrich et al. 2014). The unit's small scale (Hanif et al. 2016) and a compact design make it portable allowing its removal from the BEV for use in buildings (Rüscher et al. 2017) as well as other applications like outdoor activities.

After having started out with a 15 kW test bench for the initial evaluation of the portable micro-CHP (pmCHP) concept and optimising the unit's design space (Hanif et al. 2016), the project is in the process of finalising a scaled-down prototype.

As the test bench hardware changed across different stages of development, the software infrastructure had to adapt with only minor modifications to maintain a high degree of usability for all members of the interdisciplinary team. For this purpose, a reusable software platform for test beds was developed along with firmware for its controller, a web-based frontend and a LabVIEW-based control station (Schöner et al. 2016).

The major contribution of this publication is the detailed description of key concepts and the internal workings along with implementation details behind this platform's service infrastructure and how they help to increase its sustainability. The following pages give a short overview of the system's architecture in Sect. 3, explaining the requirements imposed on the system and the services involved along with their internal and external communication interfaces. A more detailed description of three key services for test bed configuration, test run replay and balance group evaluation follows in Sect. 4, detailing internal structures and processes. The paper concludes with Sect. 5, providing a summary of what has already been achieved and an outlook over topics for potential future research.

2 Related Work

The basis for the test bench infrastructure presented in this paper was published in Schöner et al. (2016), with a focus on usability and the interdisciplinarity of its usage. It provides an overview of the entire test bed, while a more detailed description of key components and implementation details is presented in this paper. The concept of employing pmCHP units as range extenders in BEV and buildings was published in Minnrich et al. (2014), with Rüscher et al. (2015) further investigating the scalability of the concept for use in buildings.

Like the evolution of the hardware for drivetrain testing—e.g. comparing textbooks from Hederer (1999), Paulweber and Lebert (2014)—the hard- and software to control and monitor them evolved as well. An example of this can be found in Schupbach and Balda (2002), presenting a test bed centred around a single-board hard- and software bundle for modelling, monitoring and control, bringing simplicity, albeit at a potential loss of hardware flexibility.

As a means of software-based flexibility and adaptivity, (Paulweber and Lebert 2014) suggests the use of SOA, which uses standardised interfaces to increase the speed and reduce the cost of development. In Appelrath et al. (2012), a service-oriented reference architecture for smart grid integration is proposed, in which standardised interfaces also play an important role due to diversity of devices relevant to the smart grid. This becomes especially important to the development process, once prototype status is reached and the integration of the pmCHP into smart homes and other scenarios moves into focus, requiring support for standards like e.g. IEC 61850-7-420 (Cleveland 2008).

Meanwhile another factor to be considered especially in research test benches for sustainable technologies, is the sustainability of the test bench infrastructure itself. The definition of the term 'Green IT' covers a wide range of topics related to the environmental impact of IT, but also including social and economic factors (Murugesan 2008). A differentiation is made in Calero and Piattini (2015) between 'Green in IT', which aims to reduce the environmental footprint of IT itself and 'Green by IT', evaluating means by which IT can become a tool to reduce environmental impacts from other sources. The considerations in this paper therefore regard sustainability as a goal to be achieved by creating a reusable IT infrastructure —making the soft- and hardware itself more sustainable—and by using it to reduce emissions of the actual test bench.

3 System Architecture

The first iteration of the test bench consisted of the 15 kW unit which had to be controlled manually via a switchboard directly connected to the respective actuators and with sensor readings taken by a data logger. While this was already sufficient for running many of the initial test scenarios, this setup only allowed for offline analysis of test runs with only few live readings being available via gauges.

Although some machine failures could be prevented by experienced personnel due to noise levels and several fail-safe mechanisms, fault analysis relying on measurements was only possible post-mortem, making reliable test bench operation by inexperienced personnel (e.g. students) infeasible. Due to data being recorded offline in CSV format and manual actuator operation, the immediate effects of minor alterations could also only be inspected after completing a test run, potentially leading to a high number of reruns for the same test scenario.

A set of requirements with the goal of increasing test bench efficiency and sustainability by reducing personnel requirements as well as decreasing the overall number of scenario reruns is therefore presented in the following subsection.

3.1 Requirements

To increase test bench usability, one of the main requirements is to provide near real-time sensor readings to its operators (RQ1) as well as the capability to—automatically as well as manually—transmit commands to actuators (RQ2). As an additional safety measure, the system should be able to monitor certain metrics for thresholds to prevent emergency situations by signalling the operator on exceedance as well as enacting predefined procedures in case of critical exceedance (RQ3). These measures are intended to allow safe and reliable test bench operation e.g. for students, without permanent supervision by experienced personnel, and run under near real-time conditions. All tasks with hard real-time requirements, like engine control, must be handled by external components (e.g. the controller boards (Schöner et al. 2016) (see Fig. 1)), which are not in the scope of this paper.

Since unnecessary reruns of test scenarios result in additional wear on test bench components as well as CO_2 emissions, the system needs to constantly record every sensor reading and actuator command (RQ4) to be able to replay them in a manner similar to performing a live test run (RQ5). This includes the capability to skip through and pause the replay as well as re-evaluating correction formulae—used to compensate for different tolerances and characteristics of sensors—and balance groups—collecting data from multiple sensors to compute balances between inputs and outputs—to avoid reruns due to erroneous post-processing of raw sensor data.

The configuration of balance groups, sensors and actuators, along with their characteristics and locations in the test bench, due to its experimental nature, is subject to frequent changes. Therefore, the system must be easily reconfigurable to reflect these changes without interfering with its capability to replay older recordings or requiring extensive knowledge of programming (RQ6).

In addition to this configuration level adaptivity, the software architecture of the system should impose a high level of modularity, allowing for individual components to be replaced, removed or added as work on the project progresses (RQ7). This is especially important when progressing from one stage of development to the

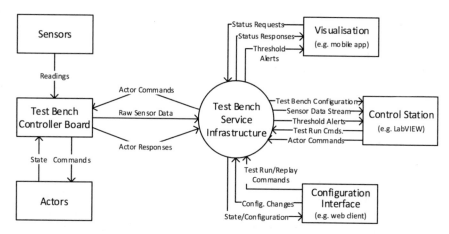

Fig. 1 System context diagram showing an overview of the test bench environment with all external components surrounding the service infrastructure

next, as this usually incurs switching to a completely new test bench, which would otherwise induce time intensive redevelopment of complex software components, in turn decreasing the sustainability of the software. Combined with the reconfigurability of the system, its modularity also allows its continued use in the long term e.g. for practical courses or student projects, giving the entire test bench a sustainable perspective even beyond the current research project.

3.2 System Context

Before going further into detail on the software components in Sect. 3.3, the system context diagram in Fig. 1 provides an overview of the test bench environment. It can be divided into test benches—consisting of the controller boards, sensors, actuators and the machine itself—and three different categories of front-end applications (control station, visualization and configuration interface) with the service infrastructure acting as mediator and containing the business logic.

The *controller boards* form the boundary between the physical sensors/actuators and the IT infrastructure and provide a basic, binary, TCP/IP-based interface. A client—e.g. test bench services—can initiate a connection to a board, creating a command and a data channel, similar to conventional FTP. While the command channel is used to transmit actuator commands (see RQ2) and controller-specific settings (e.g. sample rate) from the client to the test bench, the data channel transports a stream of sensor readings from the test bench to the client (see RQ1).

The *control station* must be provided with the test bench configuration and a constant stream of processed and enriched sensor data (see RQ1) and expects to be able to transmit commands for actuators and initiate and stop test runs (see RQ2). Like the *visualization*, it must also be notified of threshold alerts to keep the operator informed about potential alerts (see RQ3), although the visualization only serves informational purposes and requests status updates on its own without any control capabilities. The *configuration interface*—a web-based application in this case—provides the operator with a graphical front-end reflecting the current state and setup of the test bench and allowing changes to be made to both (see RQ6).

3.3   Service Architecture

Internally, the service infrastructure is made up of a total of eight different services (Fig. 2), whose orchestration is handled by *CurrentValues*, which also provides processed sensor and actuator readings to external clients (see RQ1). Additional core services are provided by *Persistence*—storage and retrieval of entities such as settings and measurements (see RQ4) to and from a database or similar data storage —and *Configuration*, which handles data on the sensor and actuator setup of the test bench as well as arbitrary settings for all other services (see RQ6).

All the other services are optional or at least replaceable components—e.g. the *Connection* service used to connect to the controller boards—for which drop-in replacements grant connectivity to different boards (see RQ7). The only service with direct access to *Connection*—aside from *CurrentValues*, receiving all raw sensor readings for processing—is *Actuator* (see RQ2), which offers interfaces for controlling all actuators retrieved from *Configuration* and can be disabled if the test bench is only to be monitored or no actuators are present.

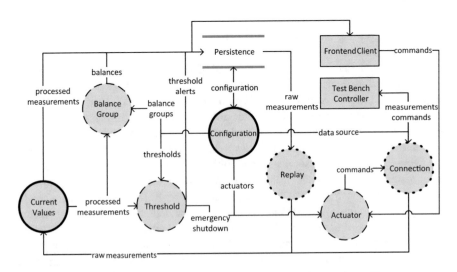

Fig. 2 Overview of the service architecture with data flows between the different services (*circles*) and external entities (*rectangles*) with core services with bold, optional with *dashed* and exchangeable with *dotted* borders

Replay is an on-demand, drop-in replacement for *Connection*, used when running replays instead of live test runs (see RQ5) and therefore also has no connection to *Actuator*. The *Balance* service retrieves balance groups from *Configuration* and monitors all processed readings from *CurrentValues* to keep them up-to-date and available to clients. *Threshold* also monitors these readings, checking them for violations of configurable hard and soft limits (see RQ3) and can be deactivated e.g. for benchmarking scenarios, to allow intentional exceedance of safety limits.

3.4 Communication Protocols and Interfaces

While internal communication as well as interfaces made available to external clients are based on RESTful APIs with JSON payloads being transmitted via HTTP, communication with the test bench controllers is handled by the controller-dependant implementation of *Connection*. This allows the use of custom binary protocols, which can be tailored to each individual controller's firmware, reducing computational requirements for communication on the embedded hardware.

The use of well documented RESTful interfaces for all other communication allows for a high degree of extensibility regarding the integration of additional services and new clients e.g. for simulation (see RQ7). This makes the entire architecture more resilient against changing requirements across all development stages from experimental to prototype test bench to actual field tests, since the software can easily be extended by adding new services. Using a service-oriented architecture also helps to limit the impact of major changes to specific services.

4 Configuration and Persistence

As mentioned in Sect. 3.3, *Configuration* is an integral part of the test bench service infrastructure, as it manages configurations for all other services, providing a key feature for the flexibility of the entire system. The *Replay* service, although not critical to the system itself, plays a key role in its sustainability, reducing the CO_2 emissions of the test bench by avoiding unnecessary reruns of scenarios. It is being supported in this task by the highly reconfigurable *Balance* service, which allows the customised computation of heat flows and summarisation in balance groups.

4.1 Configuration—Adapting Test Bench Changes

The most important feature of the *Configuration* service is the management of the test bench's sensor setup as can be seen the right part of Fig. 3 (vertical lines).

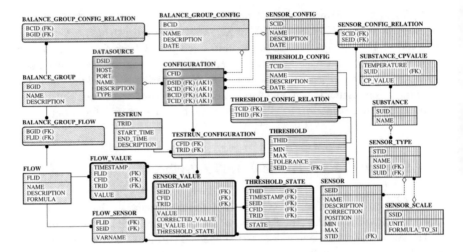

Fig. 3 ER diagram for the test bench configuration and measurement storage

Using this model, arbitrary sensors can be created by configuring new sensor types, combining a substance and its specific heat capacity at constant pressure levels (C_p) with the SI unit and custom conversion formula for the sensor.

A set of sensors can then be combined into a sensor configuration, which in turn can be used together with configurations for other components of the system and a data source (test bench controller or *Replay*) to form a configuration usable for test runs. Since each optional service comes with its own configuration model, individual configurations can be easily recombined, reused and modified between runs and components (e.g. sensor types) even across different test benches.

The combination of configurations with test runs and the capability to create partial and full copies of configurations allows multiple version of them to be kept in storage, such that previous versions connected to previous test runs always remain available. This enables the system to replay archived test runs without running into complications due to changed hardware configurations in the test bench, while still making it possible to modify parts of the virtual configurations to evaluate these modifications before performing another live test run. In the ER diagram (Fig. 3), the entities are marked with horizontal lines for *Threshold* and diagonal lines for *Balance*, with test run specific entities and readings highlighted by thicker borders.

4.2 Replay—Re-examining Recorded Test Runs

Although the data model in Fig. 3 stores raw sensor readings as well as derived values, the *Replay* service only uses the raw values when playing back a previously recorded test run. While this incurs higher computational loads than directly replaying the derived values instead, it also allows all conversion, heat flow and balance group formulae to be re-evaluated in the process, giving users the possibility to adjust and test them before performing another live run on the actual test bench. This not only saves time—thereby already increasing the sustainability of test bench operation—but also avoids unnecessary CO_2 emissions, further improving the test bench's ecological and economical sustainability.

Since all readings stored by *Persistence* contain a timestamp, *Replay* can also jump to an arbitrary point within a recorded test run, fast forwarding all readings taken before that point in time through the test bench's business logic. It is therefore possible to re-examine any time window within a recording without having to go through the entire duration of the test run first and pause execution at any given point e.g. when trying to trace back errors in balance calculations.

4.3 Balance—Keeping Track of Inputs and Outputs

Two key performance metrics in a CHP test bench are its thermal and fuel efficiency: How much of the heat generated by the ICE can be harnessed for heating purposes and how much is lost as waste heat. And how much of the chemical energy stored in its fuel is converted into useable thermal and electrical energy and how does it compare to conventional RE and electrical heating elements.

The *Balance* service therefore provides the tools to analyse these metrics by combining multiple sensors to compute energy and material flows according to fully customisable formulae. These flows can then in turn be combined to form balance groups, giving precise insights into how much energy is lost at which point of the system or how much fuel is consumed to generate which output.

This in turn allows researchers to identify potential deficiencies in the current system and further optimise it to decrease its CO_2 emissions and increase its efficiency, making the concept more sustainable and providing an eco-friendlier alternative to conventional RE in BEV.

5 Conclusions and Perspectives

In this paper, it is shown how key features of the service components of the test bed architecture presented in Schöner et al. (2016) help to increase the sustainability of a pmCHP test bench and reduce its CO_2 emissions. This is achieved by creating a highly adaptive infrastructure with few core services—e.g. *Configuration*—which provide the needed flexibility and standardised RESTful interfaces for improved interoperability. Additional services provide features to record and replay live test runs with the possibility to reconfigure balance groups and formulae between replays. An overview of the data flows between individual services was given in

Fig. 2, with a detailed ER diagram in Fig. 3 providing deeper insights into the flexibility provided by the data model behind the *Configuration* service.

While the infrastructure will remain in service in its current form for the foreseeable future, additional research is being made on further extending its feature set to increase its adaptivity as well as grant deeper insight into different usage scenarios of the pmCHP concept. The project is therefore currently evaluating the use of complex event processing to better detect, predict and prevent machine failures as well as integrate the pmCHP into a smart home or smart grid scenario.

Other avenues for further research include integration of simulation components into the system or use of recorded test runs for the calibration of external simulations, as well as the development of a prototype system for real-world testing of pmCHP units in vehicles under aspects of scalability, performance and sustainability. This includes the evaluation of pmCHP usage in fleets and car pools, but also for private use e.g. in recreational settings on board of caravans.

Acknowledgements This project was supported by the VolkswagenStiftung and the Ministry for Science and Culture of Lower Saxony (project funding number VWZN2891). We would like to thank the students who participated in the bachelor project building this service infrastructure and all our colleagues from the research focus 'Scalability of mobile Micro-CHP units' and the Institute for Engineering Design, Mechatronics and Electro Mobility (IKME) at the University of Applied Sciences and Arts Hannover for their support and the productive cooperation.

References

Appelrath H-J, Beenken P, Bischofs L, Uslar M (eds) (2012) IT-Architekturentwicklung im Smart Grid. Perspektiven für eine sichere markt- und standardbasierte Integration erneuerbarer Energien. Springer, Berlin, Heidelberg

Bundesministerium für Wirtschaft und Energie (2007) Eckpunkte von Meseberg für ein integriertes Energie- und Klimaprogramm. Die 29 Maßnahmen des IEKP. Meseberg, Deutschland. http://www.bmwi.de/Redaktion/DE/Downloads/E/eckpunkt-fuer-ein-integriertes-energie-und-klimaprogramm.pdf. Accessed 26 Apr 2017

Calero C, Piattini M (eds) (2015) Green in software engineering. Springer International Publishing, Cham

Cleveland FM (2008) IEC 61850-7-420 communications standard for distributed energy resources (DER). In: 2008 IEEE power and energy society general meeting—conversion and delivery of electrical energy in the 21st century. Energy Society General Meeting, Pittsburgh, PA, USA, 20. t- 24.07.2008, pp 1–4. IEEE. doi:10.1109/PES.2008.4596553

Hanif HI, Minnrich JP, Schmicke CRP, Rüscher H, Gusig L-O (2016) Bauraum- und gewichttechnische Untersuchung einer mobilen mikro-PCU zur Bereitstellung von Wärme, Kälte und elektrischem Strom im E-Fahrzeug. In: Steinberg, P. (ed.) Wärmemanagement des Kraftfahrzeugs X. Energiemanagement, pp 262–279. Expert Verlag, Renningen

Hederer A (1999) Elektromotorenprüfung. Computerunterstützte Elektromotorenprüfung mit klassischen und modellgestützten Verfahren für die Qualitätssicherung. Kontakt & Studium, vol 536. Expert Verlag, Renningen-Malmsheim

Kraftfahrt-Bundesamt (2016) Der Fahrzeugbestand im Überblick am 1. Januar 2016 gegenüber 1. Januar 2015. Flensburg, Germany. http://www.kba.de/DE/Statistik/Fahrzeuge/Bestand/Ueberblick/2016/2016_b_ueberblick_pdf.html. Accessed 26 Apr 2017

Minnrich JP, Rüscher H, Schmicke CRP, Gusig L-O (2014) Integration einer mikro-PCU im Thermomanagement eines Elektrofahrzeugs unter Berücksichtigung von Reichweite und Emissionen. In: Steinberg P. (ed) Wärmemanagement des Kraftfahrzeugs IX. Energiemanagement, pp 324–340. Expert Verlag, Renningen

Murugesan S (2008) Harnessing green IT. Principles practices. IT Prof. doi:10.1109/MITP.2008.10

Paulweber M, Lebert K (2014) Mess- und Prüfstandstechnik. Antriebsstrangentwicklung; Hybridisierung; Elektrifizierung. Der Fahrzeugantrieb. Springer Vieweg, Wiesbaden

Presse- und Informationsamt der Bundesregierung (2013) Deutschland soll Leitmarkt werden. Elektromobilität, Berlin, Germany. https://www.bundesregierung.de/ContentArchiv/DE/ Archiv17/Artikel/2013/05/2013-05-27-elektromobilitaet.html. Accessed 26 Apr 2017

Rüscher H, Bitner D, Saul D, Guwy A, Premier G, Gusig L-O (2017) Application scenarios for a dual use of a portable micro-CHP unit in a BEV and building. Sustainability in energy and buildings: research advances, vol 6. Submitted Mar 2017

Rüscher H, Minnrich J, Schmicke C, Gusig L (2015) Applicability and scalability of mobile mCHP units in mid-size battery electric vehicles and detached houses with different energy standards, vol 4, pp 18–23

Schmicke C, Rüscher H, Gusig L (2014) Strategies for combined use of power conditioning units in vehicles and buildings. Sustain Energ Build: Res Adv 3:31–36

Schöner D, Pump R, Schmicke C, Minnrich JP, Rüscher H, Ahlers V, Koschel A (2016) IT-Unterstützung von BHKW-Prüfständen in der angewandten Forschung. In: Mayr HC, Pinzger M (eds) INFORMATIK 2016, Klagenfurt, Österreich, 26–30. Sept. 2016. Lecture Notes in Informatics, vol 259, pp 1227–1238. Gesellschaft für Informatik, Bonn

Schupbach RM, Balda JC (2002) A versatile laboratory test bench for developing powertrains of electric vehicles. In: 2002 IEEE 56th vehicular technology conference proceedings. 2002 IEEE 56th vehicular technology conference, Vancouver, BC, Canada, 24–28 Sept. 2002, pp 1666–1670. IEEE, Piscataway, NJ. doi:10.1109/VETECF.2002.1040499

An Approach for a Comprehensive Knowledge Base for a DSS to Determine the Suitability of Open Data Business Models

Johann Schütz, Dennis Schünke, Benjamin Wagner vom Berg, Christian Linder and Frank Köster

Abstract With the technological trends and changes concepts like open data are gaining more and more public attention. Even governments and organizations recognize their data as a new raw material for economic value creation. Thus, an increasing number of organizations recognize the benefits of opening their own data. But opening their data presents the organizations with new challenges. An appropriate business model which considers the individual interests and situation of an organization while taking account on a huge amount of highly diverse influential factors is one of the most complex tasks. This paper investigates the present business models for open data and their accompanying decision factors with the main focus on smart mobility business models. Subsequently an approach for a comprehensive knowledge base, which can be used as a foundation for a decision support, is proposed.

Keywords Open data · Business models · Decision support · Sustainable mobility

J. Schütz (✉)
OFFIS e. V, Oldenburg, Germany
e-mail: johann.schuetz@uni-oldenburg.de

D. Schünke
Procedo GmbH, Oldenburg, Germany
e-mail: dennis.schuenke@uni-oldenburg.de

B. Wagner vom Berg
University of Applied Science Bremerhaven, Bremerhaven, Germany
e-mail: benjamin.wagnervomberg@hs-bremerhaven.de

C. Linder · F. Köster
German Aerospace Center (DLR), Braunschweig, Germany
e-mail: christian.linder@dlr.de

F. Köster
e-mail: Frank.Koester@dlr.de

© Springer International Publishing AG 2018
B. Otjacques et al. (eds.), *From Science to Society*, Progress in IS,
DOI 10.1007/978-3-319-65687-8_27

1 Introduction

The trend of the urbanization is accompanied by many challenges. One of these challenges results from the fact, that the demand for mobility is constantly growing while the transport infrastructure remains the same (Jaekel 2015). Therefore, to handle the future level of mobility, the existing infrastructure capacities have to be optimized. One considered key-requirement to reach this objective is the concept of open data (Gartner 2015). Thus, an increasing number of governments and organizations recognize the benefits of opening their own data (Immonen et al. 2014). For this reason, data is gaining an increasingly significant importance and is even considered as a new raw material for economic value creation (Immonen et al. 2014). But opening their data presents the organizations with new challenges. So one of the most complex tasks is to find and estimate the suitability of an appropriate business model for this kind of information or data sharing which considers the individual interests and situation of an organization while taking account on a huge amount of highly diverse influential factors.

The suitability of a business model for a specific use case is influenced by many factors (Zeleti et al. 2014). Especially the domain of open data and smart mobility provides some particular influences which have to be identified and considered for choosing and/or developing a successful business model. Furthermore, not every business model is influenced by the same factors. So the biggest challenge lies in the identification of the associations between business models and their individual influential factors. An interlinking between both can establish a knowledge base which allows an appropriate decision support for all kinds of use cases in the specific domain (Scholl 2012).

With the aim to facilitate the decision-making process and support organizations to make an appropriate choice (and therefore support a sustainable mobility in smart cities), a comprehensive knowledge base had to be build-up (Sprague and Carlson 1982). Based on the current literature this paper represents the present business models for open data and their accompanying decision factors with the main focus on smart mobility.

2 Research Objectives and Methodology

An initial research in the field of open data and smart mobility in relation to business models has shown, that within this domain several unacknowledged research questions arise. One of the key issues was the unresolved question: Which eligible domain specific business models currently exists? Or rather, which factors are pertinent for an appropriate decision-making? With this primary objective in mind it is necessary to identify the general and specific characteristics, requirements and dependencies of each business model to build up a fundamental knowledge base which contains all decision-making factors.

To identify the existing open data business models and their influential factors a literature search was carried out. Due to the importance of actuality only the years 2014–2016 were considered in this research. With the keyword-combinations: "business model" or "business case" or "business concept" or "business innovation" and "open data" or "data sharing" or "standards" or "mobility" or "smart city" were searched within a total of 11 scientific databases in the fields of computer science, business informatics and business management, such as IEEE, ACM, EBSCOhost and SpringerLink.

Based on the results of this research and the subsequent analysis several diverse open data business models and influential factors could be determined. However, a decision support requires beyond these incoherent decision-making factors an interlinking between the identified business models and their appropriate influential factors (Klein and Methlie 2009). For this reason, an interlinking matrix was developed. The results will be shown in this paper.

3 Decision-Making Knowledge Base

Decision-making factors can be divided in two essential aspects: The decision criteria or rather influential factors which represents the general conditions and requirements of the underlying use case and the options for action which in this case represents the scope of eligible business models (Sprague and Carlson 1982). Like in Fig. 1 visualized, both decision-making factors can be viewed at many different levels which are represented through the layers. But, as deeper or more detailed the viewed level/layer gets, the maintainance and inclusion of all influential factors in the decision-making process equaly becomes more complex. Figure 2 provides an example for the increasing complexity. So, for example, the influential dimension "Role" (See Fig. 3) could be broken down at least up to four deeper levels. If this example is considered as the average case of each dimension, this means on an average every influential factor could be broken down up to 720

Fig. 1 Comprehensive decision-making knowledgebase architecture

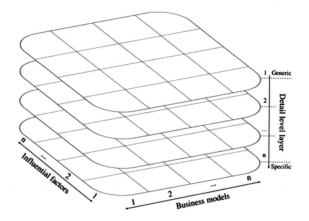

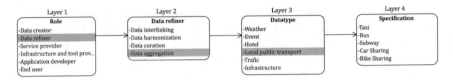

Fig. 2 An example for the considered selectiveness and granularity of influential factors

(6 possible roles * 5 possible data refiner roles * 6 possible datatypes * 6 possible specifications of datatypes) more detailed factors. Based on this assumption, all in Fig. 3 represented influential factors together could add up to over 50,000 conceivable factors.

Besides the nearly impossible applicability of the vast number of conceivable factors, they are on different layer as the currently determined business models within the proposed framework. Therefore, these factors first take place in more specific business models (Osterwalder et al. 2005; Wirtz 2010). So for the very same reason the business models as well as the influential factors can only be considered if they are on a same generic level on the same layer.

Decision criteria After every in Fig. 3 listed influential factor was identified, each of them were analyzed for its cause and effect. We determined five classes which can be clustered by their causalities:

1. Own situation: Influential dimensions of this class describe the suitability of a business model for an entity. By determining the current (or in future desired) position this dimensions enable a first estimation of the suitability of a business model for the determined position.

2. Business model (influences): Like Table 1 demonstrates, every business model has its own individual inseparable influences.

3. Business model (supported characteristics): In addition to class 2 influences, every business model has a set of supported characteristics. These can be regarded as applicable configurations within the business model concept which can be varied for a use case without a violation of the very same.

4. Additional internal influences: This class groups additional internal influential factors. Even though they have a huge impact on a business model and have to be considered, they first crystallize in a concrete use case. So they can only be considered in a later phase of the business model development-process.

5. Additional external influences: Influential dimensions of this class are additional external factors which also have to be considered. But like class 4 they first crystallize in a concrete use case. So they can only be considered in a later phase of the business model development-process too.

These five classes again can be divided in two types which lie on different layers. The first three classes lie on the first layer and therefore have to be applied within

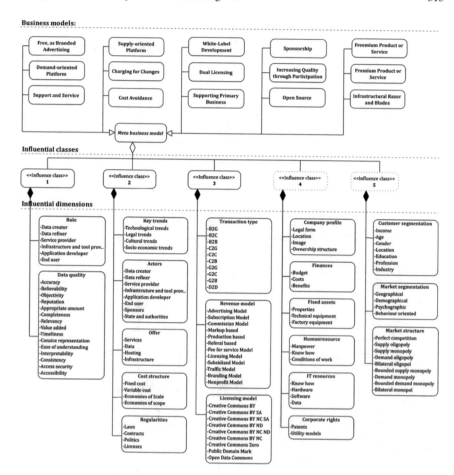

Fig. 3 Summary of all decision-making factors—influential factors as well as options for action

the generic level of a business model. While the fourth and fifth class lie on the second layer and therefore are applicable within the more specific business models. Since the fourth and fifth class need a specific use case to estimate their requirements and impact, neither one can be considered in this early step of decision making. Even if both should be considered after the underlying business model-concept was chosen.

Options for action Due to the core concept of open data to be free (OKI 2016) it is in a natural conflict of objectives with (usually) profit-orientated business models. Therefore, the definition of open data in the scope of this article is analogous to Lindman et al. (2014) extended by: *"Data, which is legally accessible through the internet in a machine-readable format. It does not have to be completely free of charge or restrictive licenses, but experimenting with the data, and running a small- scale-business should be legal"*. Under this condition we identified

Table 1 Relations between influential factors and the options for action

Influence factor	Premium Product/Service	Freemium Product/Service	Open Source	Infrastructural Razor & Blades	Demand-oriented Platform	Supply-oriented Platform	Free, as Branded Advertising	White-Label Development	Sponsorship	Dual Licensing	Support and Service	Charging for Changes	Increasing Quality through Participation	Cost Avoidance	Supporting Primary Business
Role [3][4][7][9][10][12][14][21][22][27][30][31][33]															
Data creator								•							
Data refiner					•	•	•			•		•		•	
Service provider	•	•	•	•	•	•	•		•	•	•	•	•	•	•
Infrastructure & tool provider					•	•									
Application developer		•					•	•							
End user															
Data quality [4][9][13][27][31][33]															
Accuracy															
Believability	•	•											•		
Objectivity															
Reputation	•														
Appropriate amount															
Completeness													•		
Relevancy															
Value added	•	•	•		•	•							•		
Timeliness															
Concise representation															
Ease of understanding															
Interpretability															
Consistency					•								•		
Access security															
Accessibility			•		•	•							•		
Key trends [19][24]															
Technological trends	•	•	•	•	•	•	•	•	•	•	•	•	•	•	•
Legal trends	•	•	•	•	•	•	•	•	•	•	•	•	•	•	•

(Left margin group markers: "1" spans the Data quality section; "2" spans the Key trends section.)

(continued)

Table 1 (continued)

Cultural trends	•	•	•	•	•	•	•	•	•	•	•	•	•	•	•
Socio economic trends	•	•	•	•	•	•	•	•	•	•	•	•	•	•	•
Actors [3][4][7][9][10][12][14][21][22][27][30][31][33]															
Data creator	•	•	•	•	•	•	•	•	•				•	•	•
Data refiner					•	•	•	•					•	•	
Service provider	•	•	•	•	•	•	•	•		•	•	•			•
Infrastructure & tool provider					•	•									•
Application developer			•			•		•							•
End user	•	•	•	•	•	•	•	•	•	•			•	•	•
Sponsors									•						
State and authorities						•							•		•
Offer [4][9][10][22][27][33]															
Services	•	•	•	•	•	•	•	•			•	•			
Data	•	•	•	•	•	•	•	•	•	•				•	•
Hosting					•	•									
Infrastructure					•	•									
Cost structure [19]															
Fixed cost	•	•			•	•	•	•			•	•			
Variable cost					•	•					•	•			
Quantitative benefits								•							
Economies of scope				•				•			•		•		•
Regularities [24][27][33]															
Laws	•	•										•			
Contracts					•	•	•		•	•		•	•		
Politics															
Licenses	•	•	•	•					•		•		•		
Transaction type [1][4][9][10][27][33]															
B2G							•								
B2C	•	•	•				•	•	•				•	•	•
B2B	•								•	•			•		
C2G															
C2C															
C2B														•	
G2G															
G2C															•
G2B															•
D2D															

(continued)

316 J. Schütz et al.

Table 1 (continued)

Revenue model [3][4][9][10][27][29][33]															
Advertising Model		•					•								
Subscription Model	•						•								
Commission Model																
Markup based																
Production based							•									
Referral based																
Fee for service Model	•	•		•	•	•		•			•	•				
Licensing Model			•							•						
Subsidized Model			•	•												
Traffic Model																
Branding Model							•									
Nonprofit Model		•		•	•									•	•	•
Licensing model [11][17]																
Creative Commons[a]	•	•	•	•	•	•	•	•	•	•	•	•	•	•	•	•
Public Domain Mark	•	•	•	•	•	•	•	•	•	•	•	•	•	•	•	•
Open Data Commons	•	•	•	•	•	•	•	•	•	•	•	•	•	•	•	•

[a]The same occurrence applies to all variations and versions of the creative commons licenses

15 business model-concepts. Each of these 15 substantially different concepts define a generic *template* which allows in accordance with Osterwalder et al. (2005), Wirtz (2010) a classification of concrete business model use cases. So the quasi-unlimited scope of conceivable use cases can be handled on a generic meta-level.

As the result of the previous literature research we identified during the first step 15 generic business model-concepts as well as approximately 100 possible influences within the domain of smart mobility and open data which can be grouped by similarities in 19 dimensions. Figure 3 provides an overview of all identified decision-making factors.

Interlinking However, these incoherent decision-making factors alone cannot support the decision-making process (Scholl 2012). For this reason, they have to be connected to each other. The elaborated interlinked results which are shown in Table 1 can be regarded as the first layer of the introduced framework and already be used as a foundation for a decision support.[1] But, it still needs to be explored in further research which decision-making model or method is suited best.

---

[1]Since the results of Table 1 are limited to the presented literature research, the completeness cannot be guaranteed.

4 Conclusions and Future Research Implications

Even if the built knowledge base provides a sophisticated foundation for an appropriate estimation of the suitability of an open data business model. All identified, presented and interlinked decision factors in Table 1 represent just the first generic layer of the in Fig. 1 Comprehensive decision-making introduced framework. Therefore, the characteristics of each business model in relation to each influential factor over each detail level layer has to be specified in further research like it has been exemplary done in Fig. 2. The extension of the knowledge base over all layers would allow new multiple use cases of this proposed framework which are accompanied by unacknowledged research questions:

- Even though the first generic layer of the framework in conclusion has a weak reference to the sector of smart mobility, the established knowledge base is applicable to all open data-initiatives regardless of the field. In order to close the gaps and establish a stronger correlation between the proposed framework and smart mobility the deeper layers need to be elaborated in further work.
- Every influential factor currently disposes the same priority. But it can be assumed, that based on the underlying business model-concept, some factors have a higher priority than others. Therefore, the authentic ratio between the influential factors within a business model has to be identified. Based on that, the results in Table 1 can be refined.
- The objectives of a business model can conflict an influential factor. Therefore, to ensure a logical combination or interlinking of the decision-making factors, consistent constrains and rules have to be identified.
- Which decision-making model or method is suited best has to be explored and evaluated.

References

Albinsaad H (2016) How e-commerce affects the trading system. Int J Sci Technol Res 5

Donker FW, van Loenen B (2016) Sustainable business models for public sector open data providers. Delft University of Technology

Ferro E, Osella M (2013) Eight business model archetypes for PSI re-use. In: Open data on the web workshop

Gartner (2015) Open data governance is key to building a smart city (3, September 2015). URL: https://www.gartner.com/doc/3124418/open-data-governance-key-building

Immonen A, Palviainen M, Ovaska E (2014) Requirements of an Open Data Based Business Ecosystem. In: IEEE access 2, pp 88–103. issn: 2169–3536

Jaekel M (2015) Smart City wird Realität - Wegweiser für neue Urbanitäten in der Digitalmoderne. Springer Vieweg

Kitchin R (2014) The data revolution—big data, open data, data infrastructures and their consequences

Kitchin R, Collins S, Frost D (2015) Funding models for open access repositories. The Royal Irish Academy

Kreutzer T (2014) Open COnTenT—open content—a practical guide to using creative commons licences. German Commission for UNESCO

Kuk G, Davies T (2011) The roles of agency and artifacts in assembling open data complementarities. In: Proceedings of the international conference on information systems

Laranjeiro N, Soydemir SN, Bernardino J (2015) A survey on data quality: classifying poor data. In: 2015 IEEE 21st Pacific rim international symposium on dependable computing (PRDC), pp 179–188

Latif A et al (2009) The Linked data value chain: a lightweight model for business engineers. In: 5th International conference on semantic systems (I-Semantics 2009)

Lindman J, Kinnari T, Rossi M (2014) Industrial open data: case studies of early open data entrepreneurs. In: 2014 47th Hawaii international conference on system sciences, pp 739–748

Klein M, Methlie LB (2009) Knowledge-based decision support systems with applications in business: a decision support approach, 2nd edn. Wiley

Miller P, Styles R, Heath T (2008) Open data commons—a licence for open data

OKI (2016) The open definition. URL: http://opendefinition.org/

Osterwalder A, Pigneur Y (2014) Business model generation. Wiley

Osterwalder A, Pigneur Y, Tucci CL (2005) Clarifying business models: Origins, present, and future of the concept. Commun Assoc Inf Syst 16:1–25

Poikola A, Kola P, Hintikka KA (2011) Public data—an introduction to opening information resources. Ministry of transport and communications

Ponte D (2015) Enabling an open data ecosystem: preliminary findings from the market. In: Twenty-third european conference on information systems (ECIS)

Schallmo Daniel (2013) Geschäftsmodelle erfolgreich entwickeln und implementieren. Springer, Berlin, Heidelberg

Scholl A (2012) Planung und Entscheidung: Konzepte, Modelle und Methoden einer modernen betriebswirtschaftlichen Entscheidungsanalyse. Vahlen, 2nd edn

Sprague RH, Carlson ED (1982) Building effective decision support systems. Prentice Hall, Professional Technical Reference

Stuhlberger G (2016) Geschäftsmodelle in open government data

Tucci A (2003) Internet business models and strategies text and cases—a taxonomy of business model. 2. The McGraw-Hill Companies

Välimäki, Mikko (2003) Dual licensing in open source software industry. In: Systemes 'd Information et Management 8.1, pp 63–75

Wang RY, Strong DM (1996) Beyond accuracy: what data quality means to data consumers. J Manage Inf Syst 12(4):5–33

Wirtz BW (2010) Business model management: design—Instrumente—Erfolgsfaktoren von Geschäftsmodellen. Gabler Verlag

Zeleti FA, Ojo A, Curry E (2014) Business models for the open data industry: characterization and analysis of emerging models

New Era of Fleet Management Systems for Autonomous Vehicles

Alexander Sandau and Jorge Marx Gómez

Abstract The automotive sector is influenced by several major trends, such as cost pressure, decreasing car ownership, digitalization and huge investments in technologies for electric vehicles and autonomous driving. These leads to huge challenges for automotive manufacturers and their business models. The transformation from manufacturers to mobility service providers is one suitable approach to overcome these challenges. Simultaneously, modern ICT can enhance vehicles to self-propelled, high-technical control units, that enables the driver to assign tasks to the autopilot step by step. This poses considerable challenges for the automobile branch and raises questions as "How car manufacturers can differentiate themselves in the future, if the vehicle is not being paramount?". Since the users of fully autonomous vehicles have no longer to deal with driving, the possibilities of the interior design and new value-adding services can act as a kind of demarcation. The driving time can be used for different activities as work or leisure. Empowered by this technological development the market asks for more customer centered mobility services with value-adding offers. Following these development, new intelligent fleet management systems are needed, that are capable to react on events from autonomous vehicles and providing value-added services for consumers. As main contribution, the paper introduces an agent-based approach for marketing value-added services in fleet management systems.

Keywords Mobility · Autonomous vehicles · Autonomous mobility · Fleet management · Mobility services

A. Sandau (✉) · J. Marx Gómez
Department of Business Information Systems, University of Oldenburg, Oldenburg, Germany
e-mail: alexander.sandau@uni-oldenburg.de

J. Marx Gómez
e-mail: jorge.marx.gomez@uni-oldenburg.de

© Springer International Publishing AG 2018
B. Otjacques et al. (eds.), *From Science to Society*, Progress in IS,
DOI 10.1007/978-3-319-65687-8_28

319

1   Introduction

A car will need its very own ecosystem. An independent virtual cloud ecosystem is needed
to balance the power between end-consumers, digital tech giants and traditional 'offline'
hardware companies like auto manufacturers.

(Moritz Pawelke, Global Executive for Automotive, KPMG 2017, p. 35)

The global automotive market is still booming and large parts of the world, especially emerging and developing countries, such as China and Russia, have a high demand of cars (PricewaterhouseCoopers, p. 12). The development of the automobile fulfills a significant change, from a purely technical machine to a customer- and service-oriented overall package—a digital ecosystem.

While manufacturers have already been impressed with classic features such as engine performance, acceleration times or interior comfort, a new focus has been emerged. One of the most recent places of retreat and closure—the analogue car—is threatened by electronic helpers, such as modern assistance systems from safety and infotainment sector, which are revolutionizing the vehicle of today and tomorrow. This is not an exaggeration, the private car is already coupled with the Internet and the associated, far-reaching networks, which also reached many other areas of everyday life (cf. Bolse and Heinrich 2014).

Today, modern ICT can enhance vehicles to self-propelled, high-technical control units, that enable the driver, on the one hand, to assign tasks to the autopilot step by step. Later, the vehicles are able to run independently, depending on the requirements and the intended use, while the driver can concentrate on other, usually equally, digital activities. On the other hand, completely new service and digital value-adding services in the field of Mobility-as-a-Service (MaaS Alliance 2017) will become possible.

Modern communication technology and resulting business models as alternative solutions such as rental services are the reasons for a decreasing owning rate of private cars (Canzler and Knie 2016). Furthermore, the number of low-income persons increases, making the granting of loans for car purchases more difficult (Tully 2013). In the future, automotive manufacturers can counteract the insignificance of car ownership by means of innovative sales and business models.

However, the marketing of automobiles is becoming increasingly difficult. This requires a rethinking of car manufacturers, they are no longer manufacturers of cars, they will become service providers of mobility (cf. Wagner vom Berg et al. 2017). This poses considerable challenges for the automobile groups and raises questions, as "How car manufacturers can differentiate themselves in the future, if the vehicle is not being paramount?". Since the users of fully autonomous vehicles are no longer had to deal with driving, the possibilities in the interior design and connected value-added services represent a kind of demarcation. The driving time can be used differently in the future, for work or leisure activities. This offers a great potential for value creation of manufacturers. Furthermore, the manufacturers can distinguish themselves as mobility providers with different value-added services. As soon as autonomous vehicles have reached the maturity phase, many manufacturers are planning to operate fleets of robo taxis. As a result, more intelligent mobility services occurring and replacing existing services as car sharing.

The next chapter describes the limitation of current systems. Chapter 3 introduces an agent based fleet management system model for value-added services. Chapter 4 closes with a short conclusion.

2 Limitations of Current Fleet Management Systems

Currently, the focus of automotive manufacturers lies strongly on the purely technical development of fully autonomous vehicles. In order to be able to generate future sources of income, it is necessary to gain an overview of saleable value-adding services in form of existing technologies, that are already introduced by concept studies. Current (fleet management) systems are not capable of handling personalization of vehicle preferences regarding specific customers. Examples of personalized preferences are the control of the preferred interior temperature, settings of the seat position or volume of multimedia services such as navigation and radio. In order to meet the requirements of comfort in the future, it is necessary for a customer to be recognized by the different vehicles of a fleet and automatically made adjustments in the interior. For instance, to increase the comfort of shared mobility, the seat setting, the regulation of the temperature or further comfort functions must be fully automatic, depending on the data and wishes of the customer previously defined in the system. The marketing potential is very promising if these services are offered as packages in the future. However, many concepts and studies of autonomous driving exist, but there is no implementation of value-adding services in fleet management systems yet. At this point, an information system is required that can address existing and future technologies for value-adding services via appropriate interfaces.

2.1 Current Development of Autonomous Vehicles

Currently the developing of autonomous vehicles proceeding very fast. The SAE International defined 6 levels of autonomous cars: Level 0: Automated system issues warnings but has no vehicle control; Level 1: Driver and automated system shares control over the vehicle; Level 2: The automated system takes full control of the vehicle; Level 3: The driver can safely turn their attention away from the driving tasks; Level 4: Driver attention is not required for safety; Level 5: No human intervention is required (Bartels et al. 2015).

The manufacturers of future, self-propelled cars are planning to invest more and more in the future-oriented technologies (Spickermann et al. 2013). As example,

Mercedes-Benz already sees large market shares for autonomous driving for 2020. The manufacturer plans to invest 16 billion euros in research for this division in the next years. Due to the high investments of all manufacturers, prices for vehicles will steadily increasing. This is certainly due to the extra charge for vehicles, due to the previous investments in new technologies (especially electric and autonomous mobility). For instance, the development of engines and batteries for electric mobility increases the price of the vehicles by 30–40%. This challenge must be faced by manufacturers in the future. One approach of the manufacturers is the transformation to providers of mobility services. So, cars are no longer sold to consumers, instead a temporary lease/usage of the produced vehicle is offered (Wagner vom Berg et al. 2017). Thus, ensures a continuous source of income and shifts the automotive manufacturers to service providers. But the distinguishing characteristic of the providers are changing to the distribution of specialized value-adding service bundles as sale model.

Tesla introduced lane-assistant technology and rudimentary autonomous behavior of vehicles through a software update. From the beginning, the cars are equipped with several sensors, as camera, radar, ultrasound, GPS and further. The innovative service of Tesla is to delivery additional comfort functions as a chargeable software update over the air. This was a totally new approach to provide additional functions to customers in the automotive market. Long-established market competitors are not addressing such forms of distributing new functions and features (Tietze et al. 2011). But the new upcoming usage scenarios of autonomous vehicles requesting demand tailored value-adding services from the manufacturers. They can enhance services like leasing and financing of autonomous vehicles, to increase the market penetration of manufacture-specific data sources. Furthermore, they can become important actors for innovative mobility services even in public transportation with new autonomous public transport vehicles that combine the privacy and comfort of a private car with the low costs of public transport.

The term of new and innovative mobility service provider is often linked with Uber. Uber is a role model for a whole new ICT supported and customer-oriented automotive sector. One of the biggest advantages is the penetrating use of ICT, which leads to decreasing mediating costs as more trips are mediated. Uber uses ICT to handle the whole business relation of drivers and customer pre, during and after contracting. Starting with the mediation of customer, pricing, price transparency, payment and assessment on an ICT base (Brühn and Götz 2014).

One approach for a demand tailored public transport services is initiated through the availability of (partly) autonomous vehicles. One of these projects is the CityMobil2 co-founded by the EU, where a pilot platform for automated road transport system is implemented for test purposes in several European cities. The automated transport vehicles operating without a driver in a collective mode, where they play a useful role in the transport mix. They can supply transport services (individual or collective) in areas of low or dispersed demand complementing the main public transport network (Alessandrini et al. 2015).

Literature on autonomous vehicles and their integrated value-adding services exists mainly in form of concept studies of manufacturers and suppliers of the automotive industry (Elbanhawi et al. 2015). As well as design studies of specific interior concepts. Mercedes-Benz is focusing on luxury and comfort in its "F 015 Luxury in Motion" study. Leisure and work design are dominating, a steering wheel still exists, but the interior creates the impression of a living space. The design of the Volvo concept "Concept 26" study differs by three different driving modes (Drive, Create and Relax), depending on whether the occupant drives himself or spend the time otherwise.

2.2 Current Development of Innovative Services for Autonomous Vehicles

The concepts of Connected Car (PricewaterhouseCoopers 2016) and Vehicle Cloud describing the merging of the automotive (industry) and ICT sectors that go far beyond classical assistance systems. To take in consideration, this development doesn't focus only on value-adding services or benefits of the driver, a huge (secondary) market (McKinsey&Company 2016) establishes new possibilities for Big or Smart Data and Business Intelligence in the automotive industry. The systematic evaluation and transmission of the data opens up new potentials for value creation and innovation: from the creation of detailed customer and driving profiles, tracking and monitoring, linking the vehicles to governmental authorities (tolls, road traffic offices, etc.), connection to transport infrastructure (traffic lights) and the possibility of interactions between vehicles (accident prevention, congestion avoidance) up to the establishment of intelligent transport systems (IVS) (Löw and Rothman 2015; Fluegge 2016).

The term "intelligent cloud transport" and "Internet of Things" have established the concept "vehicular cloud" (cf. Olariu et al. 2013) and the name Intelligent Transport System as a term for the software of such systems (Menguette Meneguette 2016).

Last year Volkswagen's subsidiary for commercial vehicles, MAN, launched a cloud-based digital ecosystem for commercial vehicles, under the brand "Rio". Rio contributes to a continuous data flow to the customer, providing digital value-adding services in form of entertainment, information, time-savings, comfort, but also provides a market for potential app providers. The software enables telematics to rise the driver's awareness of his possibly risky driving style or to empower the customer to lock his truck with a digital car key on the smartphone to save time-consuming handovers of keys. The long-term objective is to bind customers to this ecosystem and to reach 80 million new customers by 2025. However, this flow of data is not to be understood as a one-way connection. Constant feedback from users should steadily improve the ecosystems, so suppliers such as Continental can provide own solutions. The entire system is an open model and in a proceeding development process.

Indeed, the different perceptions of the users regarding the fault tolerance of cars and digital systems are important. While in the case of vehicles, the zero-fault tolerance in terms of breakdowns, etc., is important, users of digital systems are much more willing to accept temporary system or function errors. Errors or breakdowns of vehicles are costlier, or even dangerous to the user in the event of an accident, while errors in digital systems can be quickly and easily remedied (Over-the-Air) (KPMG 2017). Along this new business fields, the aligned companies have to be innovative too. They have to address and support this new products and services ecosystems with own solutions, for instance insurance companies. Especially the sharing economy demands new insurance products, resp. packages, which have to be developed. Further aligned businesses are the commercialization of parking spaces from private persons as offered by JustPark and many more.

As shown in these two subsections, OEMs are under huge pressure to transform their creation of value to service providing and coincidently developing autonomous vehicles. At the same time, the mobility market asking for customer centered mobility services with integrated added-value services in an open ecosystem, that fit the needs of the consumers. The following section introduces a concept for a new fleet management system, that is capable to support intelligent mobility services and provides customer centered added-value services for consumers.

3   Concept of a Fleet Management System for Autonomous Vehicles

A few companies are addressing fleet management systems for autonomous vehicles in public transport. BestMile developed an open source integration protocol for on the market available autonomous vehicles. Also, the system is capable to manage mixed fleets of autonomous and conventional vehicles. Overall, the system tries to improve the efficiency of the transportation system by intelligent algorithms. Beside these Start-ups, most traditional manufacturers are only providing basic fleet management solutions. Nevertheless, many specialized software providers of fleet management systems are capable of providing innovative systems. They are becoming indispensable partners for OEMs in the future. The Europe Director of ARI Fleet, Majk Strika, comments on heterogeneous fleet management solutions and the German market situation as follows (Inpact Mediaverlag 2016, p. 15)

> [...] There is no service provider on the German market that can serve both cars, vans, special vehicles as well as trucks with our service depth. It should also be remembered that in Germany almost 50% of the fleets are bought or financed without services. There is a great need for support around the life cycle of the vehicles.

3.1 Approach of an Agent-Based Fleet Management System

In the development of autonomous vehicles, efficient and affordable assistance systems are essential to increase the added-value of these products. The vehicles must take care of all the necessary driving conditions such as sufficient fuel state or adequate tire pressure. For these tasks, the vehicles have to have the ability to monitor itself independently. The passenger of a fully-autonomous vehicle does not have to worry about the state of operation of the vehicle. Therefore, autonomous vehicles requiring intelligent services that are able to meet these requirements and offer passengers the opportunity to pursue other activities. Agents are very suitable to react on events, such as a low fuel state or other incidents. So, new fleet management systems have to be capable to determine the most cost-effective fueling/charging station regarding the current task, or negotiating a service appointment in a car service station with given parameters.

An agent can be understood as follows: "An agent is a computer system that is located in a particular environment and is capable of performing independent actions in this environment (given) objectives."

These two examples showing the demand for intelligent systems that are capable of operating independent and autonomously. As a suitable approach, the presented system is capable to perform such tasks. As basis, common business processes (e.g. maintenance) are used as template for the new fleet management system. These processes are enhanced by the requirements and capabilities of autonomous vehicles. The resulting business processes are derived as a workflow with several tasks that have to be performed by the system. For each task one or several agents (with connected services from suppliers) can be used. Depending on the business process and the integrated data from the vehicle and customer, the workflow can be processed and optimized. At the end the product service bundle will be assed in terms of costs and environmental KPIs. Also, it is possible to apply a workflow on a group of vehicles to achieve e.g. a distribution of a fleet in a given area for providing intelligent mobility services.

3.2 Architecture of Proposed Fleet Management System

The System consists out of two different service repositories that serve as basis. The agent-based service repository and the vehicle service repository. The agent-based service repository is designed to map the workflows of the ICT services and make the technical service endpoints of suppliers accessible. The services including two different dimensions, mobility services and value-adding services. Mobility services cover routing services for door-to-door mobility, multimodal mobility or even pooling services. Further possible mobility services are the monitoring and billing of rides or predictive maintenance algorithms. The second type are value-adding services as customer tailored insurance services, concierge service (already existing in premium vehicles) or even video on demand services. The third type are internal system services, in this case an evaluation of combined services to determine costs and environmental KPIs.

The vehicle repository forms the autonomous vehicles and provides the vehicle-related KPIs and service endpoints. New innovative services can be established trough the well-known attributes of a vehicle and the consumers demands, e.g. new specialized routing algorithm can be developed, that are reducing the CO_2 emissions of a vehicle. The customer communication takes place via established interfaces, such as travel information systems or mobile applications.

In this sense, autonomous vehicles are to be depicted as physical resources in the system, which change their state in real-time. Based on these updates, mobility related offers are to be prepared, evaluated and communicated to end users. These offers include one or more service bundles.

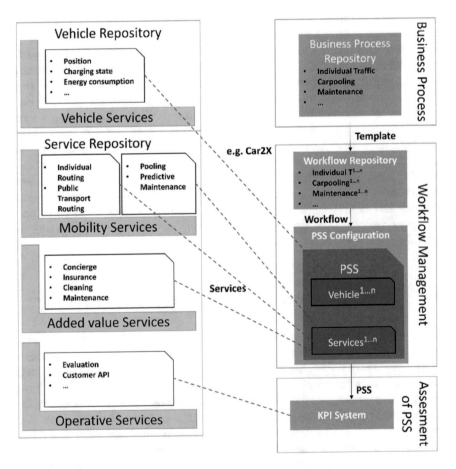

Fig. 1 Value proposition trough agent based fleet management system

The following figure shows the different repositories of the services on the left side and the integration in the creation process of Product Service Systems (PSS) bundles based on predefined business model templates and representing workflows (Fig. 1).

A prototypical implementation is possible, because the fleet management system only uses several service endpoints as routing, insurance or additional services as video on demand via cloud based APIs. The system can be hosted as cloud solution with mandators, but can also be used for municipalities and transport planners to strategical optimize fleets of autonomous vehicles in a region (traffic management). At the end, the car manufacturers only need to provide a central and secure communication interface to deliver tasks to the vehicle. Also, the system has to provide a customer API for accessing the system, e.g. as white label app and of course for separate mandators.

4 Conclusion

The automotive sector is influenced by several major trends, e.g. cost pressure, decreasing car ownership, digitalization and huge investments in technologies for electric vehicles and autonomous driving. This leads to huge challenges for automotive manufacturers. The transformation from manufacturers to provider of mobility is one suitable approach to overcome these challenges. Especially the development of autonomous vehicles can change the whole transportation system. Small political and ethical discussion are currently conducted and the technical challenges remain high.

The society and decision makers have to be integrated in the discussion to leverage the benefits of this new technology. Autonomous vehicles can improve the mobility for elders and rural areas in general, but we have to avoid an increasing amount of traffic induced by empty runs. Also, the new steering mechanisms can improve the overall mobility system and improve the access for everyone in a sustainable matter.

The proposed work is part of the research project NEMo. This an acronym (in German) for "Sustainable satisfaction of mobility demands in rural regions". The research project NEMo is developing sustainable and innovative mobility offers and supporting business models leveraged by social communities, especially in rural regions. NEMo is an inter- and transdisciplinary project and involves besides scientific participants (ICT, business information systems, social entrepreneurship, service management, law), many partners from industry (e.g. mobility providers) and public (e.g. municipalities). This provides the optimal basis for a practical oriented evaluation in rural regions in cooperation with the partners of the project.

Acknowledgements This work is part of the project "NEMo—Sustainable satisfaction of mobility demands in rural regions". The project is funded by the Ministry for Science and Culture of Lower Saxony and the Volkswagen Foundation (Volkswagen-Stiftung) through the "Niedersächsisches Vorab" grant program (grant number VWZN3122).

References

Alessandrini A et al (2015) Automated vehicles and the rethinking of mobility and cities. Transp Res Procedia, Issue 5:145–160

Bartels A, Eberle U, Knapp A (2015) System classification and glossary. H2020 AdaptIVe Project. Deliverable D2.1

Brühn T, and Götz G (2014) Die Modelle Uber und Airbnb: Unlauterer Wettbewerb oder eine neue Form der Sharing Economy?. ifo Schnelldienst 21/2014, 13 November, pp 3–27

Bolse T, Heinrich M (2014) Das Markt- und Technologieumfeld ist bereit für Connected Car. In: DMR MARKETS—Automotive. Version 2014

Canzler Weert, Knie Andreas (2016) Digitale Mobilitätsrevolution - Vom Ende des Verkehrs, wie wir ihn kannten. Oekom verlag, München

Elbanhawi M, Simic M, Jazar R (2015) In the passenger seat: investigating ride comfort measures in autonomous cars. Hg. v. RMIT University. RMIT University. Melbourne, Australia

Fluegge B (2016) Smart Mobility. 1. Wiesbaden: Springer Vieweg

Inpact Mediaverlag (2016) Zurück in die Zukunft - Der Flottenmarkt in Deutschland wird sich verändern, glaubt man bei ARI Fleet und bereitet sich mit offenen, transparenten Angeboten darauf vor. In. Mobilität der Zukunft. Technik, Trends, Innovationen. Hrsg. *Sara Karayusuf Isfahani,* Juli

KPMG (2017) KPMG's Global Automotive Executive Survey 2017

Löw M, Rothmann L (2015) Privatsphäre in smarten Interaktionsräumen? Von intelligenten Städten und der Hoffnung auf die gute Gesellschaft. In: HMD - Praxis der Wirtschaftsinformatik 52, S. 610–623

MaaS-Alliance: The MaaS Alliance. Building an open market in seamless, demand-based travel. http://maas-alliance.eu. Version: 2017

McKinsey&Company (2016) Monetizing car data—new service business opportunities to create new customer benefits. Advanced Industries, September

Meneguette R (2016) A vehicular cloud-based framework for the intelligent transport management of big cities. Int J Distrib Sens Netw 12(5)

Olariu S, Hristov T, Yan G (2013) The next paradigm shift: from vehicular networks to vehicular clouds. In: Basagni S, Conti M, Giordano S, Stojmenovic I (eds) Mobile ad hoc networking: cutting edge directions, 2nd edn . Wiley, Hoboken, NJ, USA. doi:10.1002/9781118511305. ch19

PricewaterhouseCoopers: Connected car report 2016: Opportunities, risk, and turmoil on the road to autonomous vehicles. Version: 2016

Spickermann A, Grienitz V, von der Gracht HA (2013) Heading towards a multimodal city of the future? Multi-stakeholder scenarios for urban mobility. Technological Forecasting & Social Change

Tietze F, Schiederig T, Herstatt C (2011) Firms' Transition Towards Green Product-Service-System Innovators. Working Papers/ Technologie- und Innovationsmanagement, Technische Universität Hamburg-Harburg, Issue 62

Tully C (2013) Jugend und Mobilität In: Rauschenbach, Thomas; u.a. (Hrsg.): Handbuch Jugend – Evangelische Perspektiven, S 148–154

vom Berg BW, Gómez JM, Sandau A (2017) ICT-platform to transform car dealerships to regional providers of sustainable mobility services. Interdisc J Inf Knowl Manage

Printed in the United States
By Bookmasters